S0-BBD-056

ELEMENTARY PARTICLE PHYSICS

STUDENT PHYSICS SERIES

Series Editor:
Professor R. J. Blin-Stoyle, FRS
Professor of Theoretical Physics, University of Sussex

Advisory Editors:
Professor E. R. Dobbs, University of London
Dr J. Goddard, City of London Polytechnic

The aim of the *Student Physics Series* is to cover the material required for a first degree course in physics in a series of concise, clear and readable texts. Each volume covers one of the usual sections of the physics degree course and concentrates on covering the essential features of the subject. The texts thus provide a core course in physics that all students should be expected to acquire, and to which more advanced work can be related according to ability. By concentrating on the essentials, the texts also allow a valuable perspective and accessibility not normally attainable through the more usual textbooks. The series will also include select volumes devoted to key topics at advanced undergraduate and first-year postgraduate level.

'At a time when many undergraduate textbooks illustrate inflation in poundage, both in weight and cost, an interesting countertrend is established by the introduction from Routledge of a series of small carefully written paperbacks devoted to key areas of physics. The student is offered an account of a key area of physics summarized within an attractive small paperback, and the lecturer is given the opportunity to develop a lecture treatment around this core.'—Daphne Jackson and David Hurd, *New Scientist*.

ALREADY PUBLISHED:

* **Quantum Mechanics**
P. C. W. Davies

* **Classical Mechanics**
B. P. Cowan

* **Electromagnetic Waves**
E. R. Dobbs

† **Thermal Physics**
C. B. P. Finn

* **Electricity and Magnetism**
E. R. Dobbs

* **Relativity Physics**
R. E. Turner

* **Liquids and Solids**
M. T. Sprackling

* Available in paperback only
† Available in cloth and paperback

ELEMENTARY PARTICLE PHYSICS

I. R. KENYON
Physics Department
University of Birmingham

ROUTLEDGE & KEGAN PAUL
London and New York

First published in 1987
by Routledge & Kegan Paul Ltd
11 New Fetter Lane, London EC4P 4EE

Published in the USA
by Routledge & Kegan Paul Inc.
in association with Methuen Inc.
29 West 35th Street, New York, NY 10001

Set in Times New Roman
by Cotswold Typesetting Ltd, Gloucester
and printed in Great Britain
by R. Clay & Co Ltd, Bungay, Suffolk

Library of Congress Cataloging in Publication Data
Kenyon, I. R.
Elementary particle physics.

(*Student physics series*)
Bibliography: p.
Includes index.
1. *Particles* (*Nuclear physics*) 2. *Gauge fields* (*Physics*) 3. *Hadrons. I. Title. II. Series.*
QC793.2.K46 1987 539.7'21 86–33892
ISBN 0-7102-0987-8
ISBN 0-7102-1234-8 (pbk.)

British Library CIP Data also available
ISBN 0-7102-0987-8(c)
0-7102-1234-8(p)

To Valerie

Contents

Preface

This book is intended for the use of final-year undergraduates and first-year postgraduates. The aim has been to concentrate on the 'Standard Model' and the gauge symmetries because these form the core of the subject. Leptons, quarks and forces are introduced at the beginning of the book, with a minimum of detail; then follow the experimental techniques. After this introduction the gauge theories are dealt with in order of increasing complexity. Attention is then focused on the hadrons. Deep inelastic scattering of hadrons is dealt with first, then hadron spectroscopy and finally hadron interactions. Current developments beyond the standard model appear in a last chapter.

The appendices contain mathematical detail and other material not included in the main text. These appendices cover kinematic, cross-section and decay-rate formulae; Breit-Wigner resonances; some Clebsch-Gordan coefficient tables; a table of particle properties; a set of exercises and detailed answers; and the Dirac equation. One appendix is devoted to calculating the scattering amplitudes for fermion + fermion going to fermion + fermion, which is, if anything, the 'basic' process. The appendices, apart from tabulations, are mainly intended for the postgraduate, though the interested undergraduate may also find them valuable. Up-to-date references are given at the end of the book.

Acknowledgments

I would like to thank Professor R. J. Blin-Stoyle for his encouragement and advice and for helpful criticisms of the manuscript. Also I would like to thank Professor D. C. Colley and Professor J. D. Dowell, my colleagues Dr John Garvey, Dr Martyn Corden, Dr Goronwy Jones and Dr Raymond Jones, for reading and commenting on portions of the drafts. Their generous and kindly help is very much appreciated by the author. Dr Ian Aitchison read parts of the drafts and extracted me from a number of theoretical pitfalls, for which I am most grateful. I lay claim, however, to all the errors. My thanks also go to Mr David Stonestreet of Routledge & Kegan Paul for his never-failing patience and assistance at *all* stages.

I would like to thank Margaret Baggott for her efficient preparation of the many diagrams. Finally, I would like to thank Sharon Ellis for her efficiency and speed in word-processing the text and for coping so cheerfully with the complexities of my corrections.

I acknowledge the following generous permissions to publish photographs and diagrams: from *Physical Review* and *Physical Review Letters*, granted by the American Physical Society; from *Nature*, granted by Macmillan Journals; from *Contemporary Physics* granted by Taylor & Francis; from *Reports on Progress in Physics*, granted by the Institute of Physics; from *Physics Letters*, *Nuclear Physics*, *Nuclear Instruments and Methods*, and *Physics Reports*, granted by North-Holland Physics Publishing; from *Scientific American*, granted by W. H. Freeman; from *Zeitschrift für Physik*, granted by Springer-Verlag; from *Il Nuovo Cimento*, granted by the Direttore; CERN, Geneva, Switzerland.

Chapter 1
Introduction

Elementary particle physics is the study of the ultimate constituents of matter and of their interactions. Our understanding of these interactions, as embodied in what is called the 'standard' model, has reached a degree of coherence it would have been optimistic to predict twenty years ago. The basic idea is that symmetries of nature give rise to forces; the forces are called gauge forces and the theory is called gauge theory. This is the case for the electromagnetic, the weak nuclear, the strong nuclear and the gravitational forces, i.e. for all known forces. In the case of the gravitational force the symmetry is that of space-time itself but for the other forces the symmetries refer to 'internal' spaces not directly accessible to our senses. At the range of the nuclear radius the gravitational force is negligible compared to the other forces. The other interactions are described by quantum field theories incorporating the gauge principle: the quanta are the photons, gluons, $W^{\pm}$ and Z^0 bosons. It is these quanta that are exchanged in the interactions between the material particles, the leptons and quarks. There are important unsolved problems within the standard model; however, the coverage and economy of this theory indicate that it contains essential ingredients of any more complete theory.

A principle objective of this text is to make the reader conversant with the gauge theories because these are the crux of the subject. This objective leads to departures from the more historical perspective adopted in many undergraduate texts on particle physics. If gauge theories are to be focal points then it is necessary to discuss the Dirac equation, quantum electrodynamics (the template for the other gauge theories) and also symmetries at an early stage. This essential material must of course be presented at an appropriate level.

Chapter 2 sets the stage with the discovery and properties of the material particles, i.e. the quarks and leptons. Whilst leptons are observed directly, the quarks are always bound inside strongly interacting particles (called hadrons) such as protons, π-mesons, etc. Some mention is therefore made at this stage of the families of hadrons and of the 'flavour' quantum numbers that label them.

The experimental methods of particle physics are outlined in Chapter 3 with discussion of both accelerators and detectors. This material introduces the reader to the techniques which have been useful in studying the phenomena of the very small. Chapter 4 contains an outline of the theory of quantum electrodynamics, starting with the Dirac equation and proceeding to the rates of electromagnetic processes. This chapter ends with an explanation of the way that gauge invariance (the independence of the results of physical measurement to local symmetry transformations) leads to a gauge force. In this case the force is the electromagnetic force.

Symmetries and conservation laws are dealt with more broadly in Chapter 5. The treatment provides the framework for discussing the more complex symmetries that underlie the strong force and the unification of the weak and electromagnetic forces. Discrete as well as continuous symmetries are discussed; certain discrete symmetries (parity and charge conjugation) are particularly important in understanding the weak interaction.

After this ground work the next three chapters tackle, in order of increasing theoretical complexity, the strong force and electroweak unification. Chapter 6 contains a discussion of colour symmetry and from this are deduced the properties of the force between quarks (and antiquarks). It is now possible to appreciate how the quark content of stable hadrons relates to the colour symmetry. Chapter 6 includes an account of quantum chromodynamics, the quantum field theory of the strong interaction. The reader will discover why the strong force becomes progressively weaker at shorter ranges, yet saturates at the 'long' range of the order of 1 fermi (10^{-15} m). It is the scale set by the strong force which is the key to the existence of the accidental flavour symmetries so beloved of particle physicists in the 1960s. The quanta for the strong force are called *gluons* (analogous to the photons of electromagnetism). Collectively the gluons, quarks and antiquarks make up the strongly interacting fundamental particles and are called *partons*.

This leaves the weak nuclear force, which is responsible for β-decay and for the generation of power by the sun. Chapter 7 gives an account of how our understanding of the weak force progressed from Fermi's ansatz of 1932, through the discovery of parity violation, and on to the flowering of the $V-A$ model and the realization that the weak and strong forces operate on different eigenstates of the quarks (as if two interactions picked out the linear and circularly polarized states of light). The unification of the short-range weak force and the long-range electromagnetic force was the crowning achievement of gauge theory. Glashow, Salam and Weinberg shared a Nobel prize in 1979 for the development of the theory, while Van der Meer and Rubbia shared a second Nobel prize in 1984 for completing the experimental confirmation of the theory. Chapter 8 sketches the theory using the analogy of electromagnetism as a guide, and continues with the discovery of the weak neutral-current interactions and the discovery of the heavy quanta ($W^{\pm}$ and Z^0 bosons) of the electroweak force.

At this point the gauge theories are in place and the emphasis changes. Chapters 9, 10 and 11 are devoted to describing what we know of the structure, spectra and properties of the hadrons that are built from quarks. Electroweakly interacting particles (such as electrons or neutrinos) make ideal probes of hadron structure. This work is described in Chapter 9. The results of these experiments concern the momentum and number distribution of the partons (quarks, antiquarks and even, by inference, gluons) inside hadrons. They confirm in detail the ideas expressed earlier in Chapters 2 and 6. In Chapter 10 the interest shifts to the spectrum of hadrons and their properties, more especially the static properties. It is explained how the hadron spectrum results from the properties of the strong force and the existence of several distinguishable 'flavours' of quark. Systematic methods of measuring the quantum numbers of hadrons are reviewed as part of the programme. Lastly, in Chapter 11 the interactions of hadrons are covered, a many-faceted subject. Some interactions exhibit clear features of an underlying parton + parton → parton + parton process in that each of the outgoing partons fragments into a distinctive collimated 'jet' of hadrons. Direct evidence for parton-parton collisions is less obvious in the bulk of interactions. Some understanding of the general features of interactions can be obtained from geometric arguments based on the size of the hadrons. One other subset of hadronic processes is also quite well understood: processes in which a 'virtual' hadron is exchanged.

Finally, Chapter 12 looks to the future. The standard model has a number of well-known difficulties—for example, the fact that there are about twenty arbitrary fundamental constants. Much effort has gone into attempts to unify the description. Experiments to detect the decay of protons and obtain better limits on the neutrino masses have an important bearing on the validity of such theories. Eventually the aim would be to unify all forces, including gravitation. It will be interesting to see how successful we shall be in carrying unification towards its logical conclusion of one force and a minimum of constants.

One further point about the content of the text needs to be made. Many of the simple and important processes we study are of the generic type 'fermion + fermion → fermion + fermion'. We note $e^+ + e^- \rightarrow$ quark + antiquark in e^+e^- annihilation to hadrons, the reverse in the Drell-Yan process; 'quark + quark → quark + quark' in jet production; 'neutrino + quark (lepton) → neutrino + quark (lepton)' in weak neutral-current processes; 'lepton + quark → lepton + quark' in electroweak probing of hadron structure. The author has therefore chosen to carry through in the appendices the details of the calculation of these processes; the results are applied systematically in the text with one notation throughout. The student needing the basic results will find them in the main text while those wishing to go more deeply into matters have that option available in the appendices.

1.1 Units and notation

For reasons of conciseness particle physicists have adopted a system of units which differs from the SI units. The basic difference is to set the values of the velocity of light (c) and of Planck's constant divided by 2π ($\hbar$) to unity. Then $\hbar$, c and their products are all dimensionless and equal to unity. To see what this means we write $\hbar$ and $\hbar c$ in the usual units

$$\hbar = 6{\cdot}582 \times 10^{-25} \text{ GeV s}$$
$$\hbar c = 1{\cdot}973 \times 10^{-16} \text{ GeV m.}$$

Then with the new units

$$6{\cdot}582 \times 10^{-25} \text{ GeV s} = 1$$
$$1{\cdot}973 \times 10^{-16} \text{ GeV m} = 1.$$

We are at liberty to pick *one* more unit, the second or the metre or the energy unit GeV; the other units are fixed by the above relations. The usual choice is to measure energies in GeV. Then the unit of time is $6{\cdot}582 \times 10^{-25}$ s and the unit of length is $1{\cdot}973 \times 10^{-16}$ m. Referred to the units inherited from nuclear physics the new unit of length is 0·197 fermis (fm) and the unit of area is 0·389 millibarns (mb).

In making practical calculations the method is to set $\hbar = c = 1$ and to measure energies in GeV. Any resulting lifetimes can then be converted to seconds by multiplying by $6{\cdot}582 \times 10^{-25}$ and any cross-sections can be converted to millibarns by multiplying by 0·389. When expressing the coupling strengths of interactions the Heaviside-Lorentz choice is made, which sets $\mu_0 = \varepsilon_0 = 1$. The fine-structure constant, α, which is $e^2/(4\pi\hbar c\varepsilon_0)$ in SI units, reduces to $e^2/4\pi$ and retains the numerical value 1/137. To take an example, the cross-section for $e^+ + e^- \to \mu^+ + \mu^-$ given in Chapter 2 is

$$\sigma = \pi\alpha^2/3E^2$$

where E is the beam CM energy in GeV. Numerically

$$\sigma = 0{\cdot}558 \times 10^{-4}/[E(\text{in GeV})]^2$$

and, converting to millibarns,

$$\sigma = 0{\cdot}558 \times 0{\cdot}389 \times 10^{-4}/[E(\text{in GeV})]^2 \text{ mb}$$
$$= 21{\cdot}7/[E(\text{in GeV})]^2 \text{ nb}.$$

With energies measured in GeV and c set to unity it follows that the momentum unit, GeV/c, and the mass unit, GeV/c^2, both collapse to GeV. We no longer write the energy equation as

$$E^2 = p^2c^2 + m^2c^4$$

but more briefly as

$$E^2 = p^2 + m^2.$$

Four-vectors in space-time are written $x \equiv (t, \mathbf{x})$, where t is time and $\mathbf{x}$ is the vector position. Four-momenta are written $p \equiv (E, \mathbf{p})$, where E is the energy and $\mathbf{p}$ is the three-momentum. Greek subscripts are used to denote components of four-vectors (0, 1, 2, 3) and Roman subscripts to denote the components of three-vectors (1, 2, 3). We also use repeated subscripts to indicate summation. Thus the scalar product of special relativity is

$$x \cdot p = x_\mu p_\mu$$
$$= x_0 p_0 - x_1 p_1 - x_2 p_2 - x_3 p_3$$
$$= tE - \mathbf{x} \cdot \mathbf{p}$$

and

$$\mathbf{x} \cdot \mathbf{p} = x_j p_j$$
$$= x_1 p_1 + x_2 p_2 + x_3 p_3.$$

We need to examine how a plane wave of frequency ν and wavelength λ appears when these conventions are applied. Initially in SI units

$$\psi(t, x) = \exp[i(kx - \omega t)]$$

where $\omega = 2\pi\nu$ and $k = 2\pi/\lambda$.

Planck's relation and de Broglie's relation make connections with energy and momentum,

$$E = \hbar\omega \text{ and } p = \hbar k.$$

Then in the new units ($\hbar = c = 1$),

$$\psi(t, x) = \exp[i(px - Et)].$$

Extending this to three space dimensions

$$\psi(t, \mathbf{x}) = \exp[-ip_\mu x_\mu].$$

The derivative $\partial/\partial x_\mu$ (also written ∂_μ) is a four-vector with components $\partial/\partial t$, $-\partial/\partial x_1$, $-\partial/\partial x_2$ and $-\partial/\partial x_3$. In SI units the energy and momentum operators are

$$+i\hbar\partial/\partial t \text{ and } -i\hbar\partial/\partial x_i$$

which now combine as a four-momentum operator

$$i\partial/\partial x_\mu \text{ or } i\partial_\mu.$$

Operating on the plane wave gives

$$i\partial_\mu\psi(t, \mathbf{x}) = p_\mu\psi(t, \mathbf{x})$$

so, as expected, it has four-momentum $p_\mu \equiv (E, \mathbf{p})$.

Chapter 2
Leptons, quarks and forces

It is well-known that matter consists of atoms and that the atom comprises a nucleus surrounded by a cloud of electrons. In turn the nucleus is made up from neutrons and protons, and we shall learn that these nucleons are themselves complex structures made from quarks. Such quarks together with the leptons, of which the electron is the best known example, appear to be the ultimate constituents of matter. Leptons and quarks are all point-like, and have half-integral spin.

The evidence for such substructure (or lack of it) comes from two types of experiment. One is spectroscopy, where broadly the experimenter has evidence for excited states. If excited states are observed then the object being studied is composite. Atoms, for example, have excited states because their component electrons can undergo transitions to unoccupied quantum levels. Similarly, nuclei show excited states and spectra. The second type of experiment is the scattering experiment, in which, for example, high-energy electrons are targetted on to hydrogen or deuterium. Scattering can be elastic

$$e^- + p \rightarrow e^- + p$$

or inelastic

$$e^- + d \rightarrow e^- + n + p.$$

Both scattering techniques can give information about the target structure. If the target is composite and the beam energy exceeds the binding energy the target can be fragmented into its components. The inelastic process shows that a deuteron is made from a proton and a neutron. By contrast, for elastic scattering the quantity measured is the angular distribution of the scattered beam particles. The angular distribution depends on the force between beam and target and on their sizes. A familiar example is Coulomb (Rutherford) scattering of point charges; this varies with the scattering angle, θ, like $\text{cosec}^4(\theta/2)$. If the target has finite size, and is therefore composite, the scattering is the coherent sum of contributions from its whole volume; this will be diffractive. The observed angular distribution will be the product of the two factors, a diffractive forward peak and a point Coulomb scattering. It is easy in this electromagnetic example to factor out the $\text{cosec}^4(\theta/2)$ component and see whether any diffraction is apparent. Electron scattering from a proton does show a diffractive forward peak corresponding to a radius $\sim 10^{-15}$ m. The proton and neutron are thus composite.

In this chapter the simplest evidence for the existence of the quarks and leptons is described. Quarks and leptons are the fundamental fermions, i.e. they have half-integral spin and obey Fermi-Dirac statistics. They are the stuff from which matter is created and are called *material* particles. We shall also meet other fundamental particles called *field* particles, which are carriers of the forces between material particles. Of these the best known example is the photon, the carrier of the electromagnetic force. All the field particles are *bosons*, and obey Bose-Einstein statistics. They are just as fundamental as the quarks and leptons. The forces through which the material particles interact will also be discussed, commencing with a few introductory remarks here.

The electromagnetic and gravitational forces, of infinite range, are well known. There are also two very short-range forces which have only been accessible to experiment this century: the weak and the strong nuclear forces. The weak nuclear force is responsible for the β-decay of nuclei and for steps in the cycle of processes by which the sun generates energy (such as $p+p \rightarrow d+e^{+}+\nu_{e}$). The strong nuclear force is responsible for binding the quarks into the finite structures called *hadrons* ($\sim 10^{-15}$ m across), of which the proton and neutron are examples. Hadrons are included in our definition of particles because of their small size. Some of the hadrons are made up of three quarks, have half-integral spin and obey Fermi-Dirac statistics: these are the *baryons*. The remaining hadrons are made up of a quark-antiquark pair and are called *mesons*. These have integral spin and obey Bose-Einstein statistics. One such meson is the π-meson. No *free* quarks have ever been detected and the evidence for their existence and properties comes from the study of hadrons. Hadron structure and properties therefore play a crucial role in this presentation.

Comparing the strengths of the forces exerted between protons which just touch we have

strong	: electromagnetic	: weak	: gravitational
1	10^{-2}	10^{-7}	10^{-38}

The gravitational force is evidently extremely weak in comparison and it will generally be ignored from here on (only a few remarks will be necessary in the final chapter). All the material particles feel the weak and electromagnetic forces; what distinguishes quarks from leptons is that only the quarks feel the strong force.

2.1 The electron and its neutrino

The electron was the first of the leptons to be discovered, being observed by J. J. Thomson in 1897. It is by far the lightest of all charged particles, weighing only 0·511 MeV/c^2 and appears to have no substructure. Internal structure was searched for most recently at the Deutsche Elektron Synchrotron Laboratory (DESY) where beams of electrons (e^-) and positrons (e^+) rotating in opposite senses in a ring accelerator (PETRA) were brought into collision. Each colliding positron and electron had a momentum of 24 GeV/c. According to Heisenberg's uncertainty principle the precision with which such a beam particle can be located is at best

$$\Delta x = h/(24 \text{ GeV}/c)$$
$$= 10^{-17} \text{ m}.$$

No diffractive effects were observed in the scattering of the electrons from the positrons and from this we conclude that the electron is less than 10^{-17} m across.

It is well known that the electron, although point-like, has an intrinsic angular momentum (spin) of magnitude $\sqrt{\frac{1}{2}(\frac{1}{2}+1)}\hbar$. (In what follows we shall loosely say a spin s when we mean $\sqrt{s(s+1)}\hbar$.) The associated magnetic moment of the electron is $\boldsymbol{\mu}=g\mu_B\mathbf{s}/c$, where μ_B is the Bohr magneton $(e\hbar/2m)$ and g is the Landé factor. According to Dirac's theory of the electron g is exactly 2·0. However, the modern view of an electron pictures it as accompanied by a cloud of photons which it is continuously emitting and reabsorbing, and when this behaviour is taken into account g is increased slightly:

$$(g-2)/2=(1\,159\,652{\cdot}4\pm0{\cdot}4)\times10^{-9}.$$

This departure of g from 2·0 causes the electron's cyclotron frequency (ω_c) and spin-precession frequency (ω_s) to differ slightly and a measurement of the difference has been used to determine $(g-2)$ with high precision. An electron of rest mass m and momentum p follows a circular path in a magnetic field B of radius r given by:

$$p=erB.$$

The relativistic expression for p is $m\beta\gamma c$, where βc is the velocity and $\gamma=(1-\beta^2)^{-\frac{1}{2}}$. Then the cyclotron frequency is:

$$\omega_c=\beta c/r=eB/m\gamma.$$

The corresponding frequency at which its magnetic moment precesses about the field is

$$\omega_s=(eB/m\gamma)\{1+(g-2)\gamma/2\}.$$

Van Dyck, Schwinberg and Dehmelt (1977) trapped single electrons in an electrostatic potential well, of the form $A(r^2-2z^2)$, produced by electrodes in the form of hyperboloids of revolution around the z-axis. A magnetic field was applied parallel to the z-axis so that the electron followed a helical path in the well. Both the angular momentum and spin component of the electron are quantized. Then if the electron makes a transition involving a change of one unit of angular momentum and a simultaneous spin-flip the energy change is

$$\hbar\omega_{\mathrm{a}} = \hbar\omega_{\mathrm{c}} - \hbar\omega_{\mathrm{s}} = \mu_{\mathrm{B}} B(g-2)$$

(or $\omega_{\mathrm{a}} = \mu_{\mathrm{B}} B(g-2)$ in the new units). The authors measured this transition frequency and obtained

$$(g-2)/2 = (1\,159\,652{\cdot}41 \pm 0{\cdot}20) \times 10^{-9},$$

in excellent agreement with the prediction of quantum electrodynamics (QED).

Dirac's theory of the electron predicted the existence of an antiparticle (positron) having all the quantum numbers reversed but with a mass equal to that of the electron. This prediction is part and parcel of all modern theories and all particles are expected to have antiparticle partners. The positron was first observed by Anderson in 1933 using a cloud chamber in an applied magnetic field. Fig. 2.1, taken from Anderson's paper, shows a track which is made up of droplets condensed on ions produced along its path by a charged particle. In the magnetic field the curvature of the trajectory increases as the particle slows down. The particle lost energy in traversing the lead plate across the centre of the chamber and must therefore have moved downward. From the sense of curvature Anderson could then identify the particle as positively charged. It must also be light because it ionizes lightly and travels far in the gas. Modern determinations show that the mass and charge of the positron agree with the mass and charge (reversed) of the electron.

Electrons are emitted in β-decay and therefore feel the weak nuclear force. A simple example of β-decay is neutron decay to a proton and an electron. If the electron and proton were the only decay products they would have equal and opposite momentum in the neutron rest frame, i.e. in the laboratory. The electron energy would therefore have a unique value ($E_{\mathrm{m}} = (m_{\mathrm{n}}^2 - m_{\mathrm{p}}^2 + m_{\mathrm{e}}^2)/2m_{\mathrm{n}}$ from Appendix A, where c is set to unity). Instead, the electron energy (E_{e}) spectrum is continuous from zero to E_{m}. Pauli speculated that an undetected, massless,

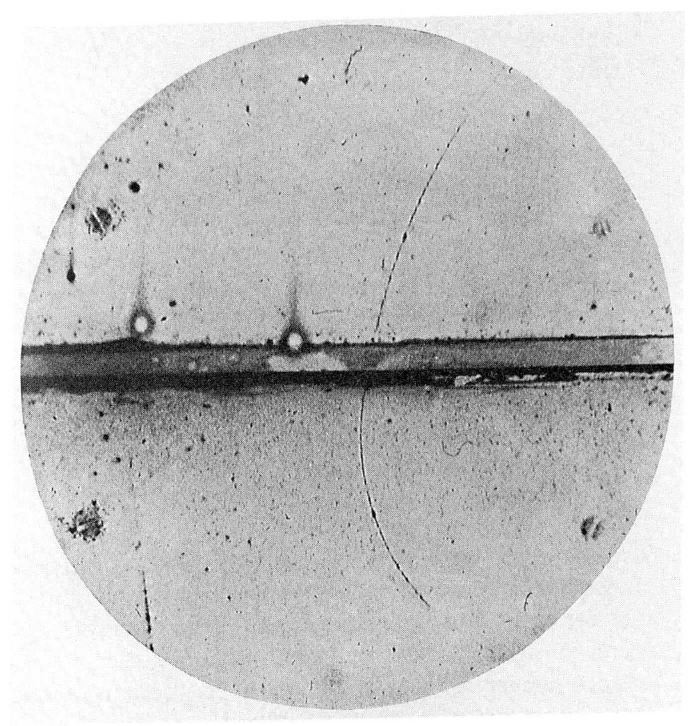

Fig. 2.1 The track of a positron in a cloud chamber (C. D. Anderson, *Phys. Rev.* **43**, 491 (1933)).

neutral spin-$\frac{1}{2}$ fermion (the electron neutrino, ν_e) was emitted in β-decay: in this view the electron energy will only equal E_m when the neutrino carries no energy. If the neutrino were to have a very small mass m_ν the maximum electron energy would be reduced; in the expression for E_m the available energy m_n would be replaced by $(m_n - m_\nu)$. The reader may like to show that E_m is reduced to $E_m - m_\nu$. Measurement of the end-point energy E_m in the decay of ^3H to ^3He plus electron, combined with precise knowledge of the masses of the charged particles involved has been used to show that the neutrino mass, m_ν, is less than 10 eV/c^2.

The antiparticle of the neutrino is also massless and neutral.

It is conventional to select the antineutrino (neutrino) to accompany the electron (positron). Then schematically the neutron decay is

$$n \to p + e^- + \bar{\nu}_e,$$

where a bar above a symbol is used to indicate an antiparticle. Any such reaction equation can be manipulated to give other equally valid reactions. For example, all particles may be changed simultaneously to antiparticles. In the present case the equation for antineutron decay into antiproton, positron and neutrino emerges:

$$\bar{n} \to \bar{p} + e^+ + \nu_e.$$

A particle can also be added to both sides to cancel antiparticles. For example, adding $n + p$ to this last equation gives

$$p \to n + e^+ + \nu_e,$$

signifying the weak decay of a proton. This last process cannot occur for free protons and neutrons because the combined mass on the right-hand side of the equation exceeds the proton mass. However, in a nucleus the neutron binding energy can be so much larger than that of the parent proton that the reaction releases energy and can take place.

It seems reasonable to inquire whether the neutrino and its antiparticle can really be distinguished—after all, they are both massless and neutral. The proof that they are different comes from studies of neutrinoless double β-decay. In principle the decay

$$^{48}Ca \to {}^{48}Ti + 2e^- + 2\bar{\nu}_e$$

is feasible but is suppressed because the intermediate nucleus ^{48}Sc is heavier than ^{48}Ca or ^{48}Ti. A lifetime (τ) of $\sim 10^{20}$ years is predicted. Now if the neutrino and antineutrino were identical the $\bar{\nu}_e$ from ^{48}Ca decay to ^{48}Sc could initiate the second step:

$$^{48}Ca \to {}^{48}Sc + e^- + \bar{\nu}_e$$

$$\bar{\nu}_e + {}^{48}Sc \to {}^{48}Ti + e^-,$$

i.e. neutrinoless double β–decay would occur with a calculated lifetime of 10^{14} years. Experimentally τ is greater than $10^{21 \cdot 7}$ years, which excludes the possibility that the neutrino and antineutrino are identical.

Neutrino-initiated reactions have a low cross-section, $\sim 10^{-40}\ E_\nu\ \mathrm{cm}^2$ for a neutrino of energy E_ν GeV incident on a nucleon. Fortunately nuclear reactors produce vast fluxes of antineutrinos from the β–decay of neutron-rich fission products ($10^{12}\ \mathrm{cm}^{-2}\ \mathrm{s}^{-1}$), a fact which Reines and Cowan (1959) made use of to initiate inverse β–decay in an external target:

$$\bar{\nu}_e + p \rightarrow n + e^+.$$

The emitted positron annihilates on an electron at rest to give two photons each with a characteristic energy of 511 KeV; and the neutron is captured radiatively on cadmium, giving a third photon some microseconds later.

2.2 The heavy leptons

The next member of the lepton family, the muon (μ- or mu-lepton) was first observed by Anderson and Neddermayer (1937), again in a cloud chamber. Muons have a unique mass about 200 times greater than that of the electron and carry an identical charge. Conversi, Pancini and Piccioni (1947) used magnetic bending to separate negative and positive muons and found that both have lifetimes of about 2 μs. Negative muons are captured in atomic orbits like electrons but being 200 times heavier their orbits are 200 times smaller. If muons interacted strongly a μ^--lepton would interact with the nucleus around which it orbits instead of decaying. Conversi's observation of equal lives for μ^+- and μ^--leptons rules out this possibility. Now the energy levels of the muonic atoms are given by the Bohr formula $-m_\mu e^4 Z^2/32n^2\pi^2$ (with $\hbar = c = 1$). By measuring the wavelength of X-rays emitted when a μ^--lepton undergoes atomic transitions the mass has been very precisely measured: $m_\mu = 105{\cdot}65932 \pm 0{\cdot}00029\ \mathrm{MeV}/c^2$.

Muons form the main component of charged cosmic rays at ground level, whilst the particles incident on the earth's atmosphere are mainly protons. The protons interact with atmospheric nuclei to give showers of other strongly interacting particles. Of these the majority are π-mesons, which have a lifetime of 26 ns and decay almost exclusively to muons:

$$\pi^+ \rightarrow \mu^+ + \nu.$$

The parentage and decay of the μ-lepton were revealed in 1947 by Powell and his colleagues using nuclear emulsions (Lattes (1947)). Fig. 2.2 shows a photomicrograph of an example of the decay sequence $\pi \rightarrow \mu \rightarrow e$ first observed by

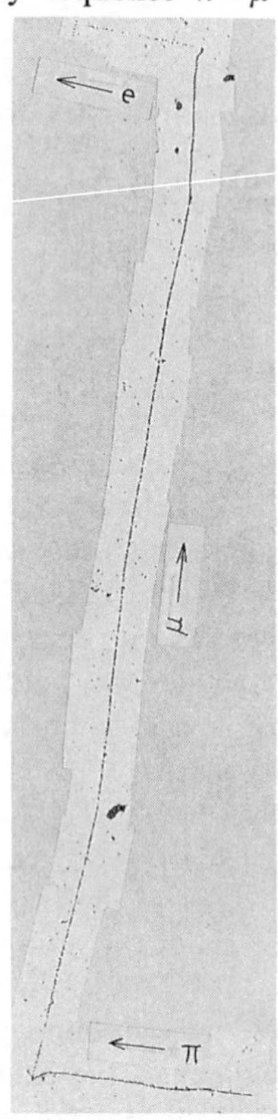

Fig. 2.2 The photomicrograph of the decay sequence $\pi \rightarrow \mu \rightarrow e$ observed in a nuclear emulsion (reprinted from *Nature* **163**, 47 (1949)).

Powell. In the emulsion the passage of charged particles through silver halide grains makes them developable. Developing the emulsion thus leaves a trail of dark silver grains along the charged-particle tracks; elsewhere the fix washes away the undeveloped silver halide. Now, as a charged particle slows down it ionizes more heavily and it undergoes larger angular deviations in collisions with atoms. Therefore it is easy to see that the μ-lepton is moving upward in Fig. 2.2. The appearance of a single track from the end point of the μ-lepton track shows that only one charged particle is emitted in its decay. Similarly for the π-meson decay. Conservation of charge then requires that the charged π-meson, μ-lepton and electron carry identical charges. Powell found that when π-mesons decayed at rest the μ-lepton emitted had a unique range ($\sim 600\ \mu$m) and hence a unique energy. This observation establishes the π-meson decay as a two-body process with the μ-lepton having an unobserved neutral decay partner (recall the remarks made above concerning β–decay). The neutral must be a neutrino rather than a photon because the neutral is never observed to convert to an electron-positron pair.

The electrons emitted in muon decay have an energy spectrum which is continuous up to a maximum and this suggests a three-body decay. The two neutrals have to be neutrinos because of the absence of any conversions:

$$\mu^- \to e^- + \bar{\nu} + \nu.$$

If the muon were simply a heavy electron it would decay electromagnetically:

$$\mu^- \to e^- + \gamma.$$

This decay has never been observed and an experimental limit has been placed on the *branching fraction* of $1{\cdot}7 \times 10^{-10}$, i.e. the fraction of μ-leptons decaying to photon plus electron is less than $1{\cdot}7 \times 10^{-10}$. Evidently the muon differs from the electron in kind and it is reasonable to expect a new type of neutrino also. Danby *et al.* (1962) demonstrated that the neutrinos produced in π-meson decay were distinct from electron neutrinos. These experimenters fired 15 GeV protons from the Brookhaven synchrotron on to a target to produce π-mesons which were allowed to decay in flight. Daughter neutrinos were filtered from other particles by a 13·5 m wall of steel. All other particles interact electromagnetically or strongly and are absorbed, leaving the neutrinos. Behind this wall was positioned a 10 tonne gas-filled spark chamber. Very infrequent neutrino interactions occurred and from them emerged the characteristically deep penetrating tracks of μ-leptons but no electrons. Thus the reaction

$$\nu_\mu + \text{nucleus} \rightarrow \mu^- + \text{other}$$

is seen but not

$$\nu_\mu + \text{nucleus} \rightarrow e^- + \text{other}.$$

It is necessary, therefore, to distinguish the muon neutrino (which accompanies the muon in π-meson decay) from the electron neutrino. We will use the notation ν_μ and ν_e from here onwards.

A common feature of all processes that involve leptons is that they conserve the net number of leptons, i.e. the number of leptons minus the number of antileptons. Because the muon and its neutrino are distinct from the electron and its neutrino there must be separate conservation of the two species:

$$\Sigma n_e = \text{constant},$$
$$\Sigma n_\mu = \text{constant},$$

where the values of n_e and n_μ are assigned as follows:

		e^-	ν_e	e^+	$\bar{\nu}_e$	μ^-	ν_μ	ν^+	$\bar{\nu}_\mu$
n_e	=	+1	+1	−1	−1	0	0	0	0
n_μ	=	0	0	0	0	+1	+1	−1	−1

The conservation laws explain why two neutrinos accompany the electron from μ-lepton decay. Putting in the labels the decay is seen to be

$$\mu^- \rightarrow \nu_\mu + e^- + \bar{\nu}_e.$$

A limit for the muon-neutrino mass can be obtained from measuring the momentum, q_μ, of the muon emitted from

π-meson decay at rest. Momentum and energy conservation require that

$$q_\nu = q_\mu$$

and

$$E_\nu + E_\mu = M_\pi.$$

From these equations we obtain

$$m_\nu^2 = m_\pi^2 + m_\mu^2 - 2m_\pi \sqrt{(q^2 + m_\mu^2)}.$$

The main uncertainty in m_ν arises from the determination of m_π and it is only possible to say that m_ν is less than 250 KeV/c^2.

The atomic spectra of muonic atoms show hyperfine splitting so that the muon has spin $\frac{1}{2}$ like the electron. It has the same charge as the electron. Thus in many ways it behaves like a heavy electron. Its magnetic moment can therefore be calculated by quantum electrodynamics. For the muon

$$(g-2)/2 = [1\,165\,920{\cdot}2 \pm 2{\cdot}0] \times 10^{-9}.$$

A series of measurements of the muon magnetic moment have been made at CERN (Centre Europeén de Recherche Nucleaire—the European centre for particle physics research). Farley and Picasso (1979) have reviewed the experiments which gave

$$(g-2)/2 = (1\,165\,924 \pm 9) \times 10^{-9}.$$

The very precise agreement between experiment and theory for both the muon and electron magnetic moments shows that our picture of the charged leptons as point-like fermions of spin $\frac{1}{2}$ and our understanding of their electromagnetic interactions are on a very firm basis.

Martin Perl and his colleagues (1975) at the Stanford Linear Accelerator Laboratory (SLAC) made a successful search for a yet heavier lepton. They studied lepton production from the collisions of beams of electrons and positrons rotating in opposite senses in an electron–positron ring accelerator (SPEAR). Some collisions were observed to produce μ^+e^- or μ^-e^+ lepton pairs and no other charged particles. The only tenable interpretation of such observations is that a heavy lepton and its antiparticle are produced with one subsequently decaying to an electron, while its partner decays to a muon. Giving the new lepton the name of τ-lepton the reaction chain can be written as

$$e^+ + e^- \rightarrow \tau^+ + \tau^-$$

followed by weak decays

$$\tau^+ \rightarrow \mu^+ + \nu_\mu + \bar{\nu}_\tau$$
$$\tau^- \rightarrow e^- + \bar{\nu}_e + \nu_\tau.$$

The absence of τ-lepton decays of the form

$$\tau^- \nrightarrow \mu^- + \mu^+ + \mu^-$$

or

$$\tau^- \nrightarrow e^- + e^+ + e^-$$

is consistent with the τ-lepton being a distinct species from either the muon or electron. Notice that the production and decay equations given above do conserve each lepton species. τ-lepton production is only possible when the centre-of-mass (CM) energy of the electron plus positron exceeds $2m_\tau$. Measurement of this threshold energy yields $m_\tau = 1\,784{\cdot}2 \pm 3{\cdot}2$ MeV/c^2, which makes the τ-lepton seventeen times heavier than the muon. A determination of the τ-neutrino mass has been made by measuring the maximum energy of the charged leptons emitted from τ-decay. The τ-neutrino mass is found to be consistent with zero within a large experimental uncertainty (<70 MeV/c^2).

To summarize, there are three distinct species of leptons, all being point-like spin-$\frac{1}{2}$ fermions which interact weakly and electromagnetically. There is separate conservation of each lepton species (e, μ, τ). Recent searches have not revealed any heavier leptons with masses less than 20 GeV/c^2.

2.3 The quark model

The evidence for quarks is not direct like that for leptons because no free quarks have ever been observed. However, the indirect evidence that strongly interacting particles (hadrons) are made up from spin-$\frac{1}{2}$ point-like fermions (quarks) appears overwhelming.

The hadrons have a rich spectrum and we need at this early stage to describe their classification briefly. There are obvious sets of hadrons which besides having identical spin-parity, have nearly identical strong interactions, and closely similar masses. The neutron and proton form one set. The π-mesons form another: $\pi^+(140)$, $\pi^0(135)$ and $\pi^-(140)$, where the number refers to the mass in MeV/c^2. Such sets are regarded as made up of the charge substates of one type of hadron. They are classified using a quantum number called the strong isospin, which has the symbol I. Strong isospin is a vector quantity behaving formally like spin. The nucleon is assigned a strong isospin of $\frac{1}{2}$, with proton and neutron substates. These substates have isospin up or down ($I_3 = \pm\frac{1}{2}$) relative to an axis in an internal isospin space; a space not connected with our usual space-time. The π-meson has unit strong isospin with three substates ($I_3 = +1, 0, -1$). Strong isospin is conserved in strong interactions. A simple example concerns the non-observation of the process:

$$\begin{array}{lcccc} & \mathrm{d} + & \mathrm{d} \not\rightarrow & \mathrm{He}^4 + & \pi^0 \\ \text{isospin} = & 0 & 0 & 0 & 1 \end{array}$$

The d and He^4 nuclei contain np and nnpp in the simplest $I = 0$ states. Vectorial addition shows that the initial and final isospins do not balance and so the reaction may not occur. The charge independence of nuclear interactions is well supported by results from nuclear physics. For example, the nuclei ^{14}O, ^{14}N and ^{14}C consist of similar cores with respectively pp, pn and nn outside. The energy levels of the O^+ states would be very different if the pp, pn and nn forces were different. In fact when allowance is made for the small electromagnetic mass differences, the energies match closely.

The 1950s saw the advent of cyclic accelerators delivering proton beams of sufficient energy to produce more massive types of hadrons called strange particles. They are termed strange because they are produced *copiously* in strong interactions yet decay *weakly*. A typical production and decay sequence is:

$$\pi^- + \text{p} \rightarrow \text{K}^0(498) + \Lambda^0(1116)$$
$$\text{K}^0(498) \rightarrow \pi^+ + \pi^- \qquad \tau = 0{\cdot}89 \times 10^{-10}\ \text{s}$$
$$\Lambda^0(1116) \rightarrow \text{p} + \pi^- \qquad \tau = 2{\cdot}63 \times 10^{-10}\ \text{s},$$

where the $\Lambda^0(1116)$ is a neutral strange baryon and the $\text{K}^0(498)$ is a neutral strange meson. The lifetimes are to be compared with the typical time taken by a strong-interaction process of $\sim 10^{-23}$ s. The paradox was resolved once it was realized that there is a new quantum number involved which we call 'strangeness', which must be conserved in strong interactions. It is given the symbol S. The assignments of strangeness for the particles so far mentioned are $S=0$ for π-mesons and nucleons, $S(\Lambda^0) = -1$ and $S(\text{K}^0) = +1$. S is taken to be an additive quantum number so that for $\text{K}^0 + \Lambda^0$ we get zero. In strong processes strangeness is conserved, so it is possible to produce strange particles in association, e.g. $\text{K}^0 + \Lambda^0$. However, a strong decay of $\Lambda^0(1116)$ is not possible because the strangeness changes in the decay to $\text{p} + \pi^-$. Weak processes, however, need not conserve strangeness and a change $|\Delta S|$ of unity is possible. This accounts for the slow decays of Λ^0 and K^0, despite the large energy release. Notice that the decay of $\Lambda^0(1116)$ to $\bar{\text{K}}^0(498) +$ n(939) *would* conserve strangeness but is not energetically possible. Nuclear β-decay, which is also a weak process, has $|\Delta S| = 0$.

A related quantum number called strong hypercharge is defined as

$$Y = B + S,$$

where B is the baryon number; zero for bosons and $+1$ for baryons. The antiparticle of a hadron with quantum numbers Y, B, S has itself the quantum numbers $-Y$, $-B$, $-S$. Charge (eQ) is empirically related to the third component of isospin (I_3):

$$Q = I_3 + Y/2.$$

In summary, the quantum numbers Q, I, I_3, B and S are conserved in strong interactions while in weak decays only Q and B are conserved.

The lightest baryons fall in clear multiplets when plotted as in Fig. 2.3(a) and (b) on a plot of Y against I_3. There is an octet with spin-parity $\frac{1}{2}^+$ and a decuplet of states with spin-parity $\frac{3}{2}^+$. The lightest mesons also fall into multiplets; the octets and singlets shown in Fig. 2.3(c) and 2.3(d). There is a singlet and an octet with spin-parity 0^-, and a singlet and an octet with spin-parity 1^-. Gell-Mann (1964) and Zweig (1964) suggested that all hadrons could be made from spin-$\frac{1}{2}$ fermion substructures, which Gell-Mann christened with the name quarks. These were supposed at that time to come in just three species given the names up (u), down (d) and strange (s). These quarks have the quantum numbers given in Table 2.1.

The baryons are built from three quarks and the mesons from a quark plus an antiquark. In the lowest energy state, which corresponds to the lowest mass hadrons, the quarks are expected to have zero relative orbital angular momentum; i.e. the quarks are in an *s*-state of relative motion. Then the overall spin of the hadron is determined by the alignment of the quark spins. The *s*-wave mesons have two possible spin states; namely 0 if the quark and antiquark have their spins aligned antiparallel; and 1 if the spins are aligned parallel. For the *s*-wave baryons the possibilities are also limited; the three quark spins can be arranged to give a spin of either $\frac{1}{2}$ or $\frac{3}{2}$. Referring to Fig. 2.3 we see these possibilities match exactly the spin combinations of the observed low-mass hadrons. The diagrams of Fig. 2.3 are known as Weight Diagrams, and will be discussed further in Chapters 5 and 10.

Table 2.1

	charge	I	I_3	Y	S	B
u	$+\frac{2}{3}e$	$\frac{1}{2}$	$+\frac{1}{2}$	$\frac{1}{3}$	0	$\frac{1}{3}$
d	$-\frac{1}{3}e$	$\frac{1}{2}$	$-\frac{1}{2}$	$\frac{1}{3}$	0	$\frac{1}{3}$
s	$-\frac{1}{3}e$	0	0	$-\frac{2}{3}$	-1	$\frac{1}{3}$

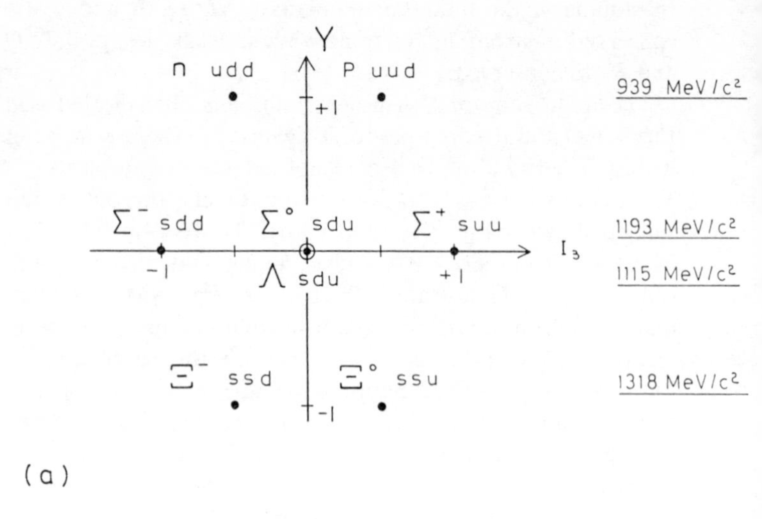

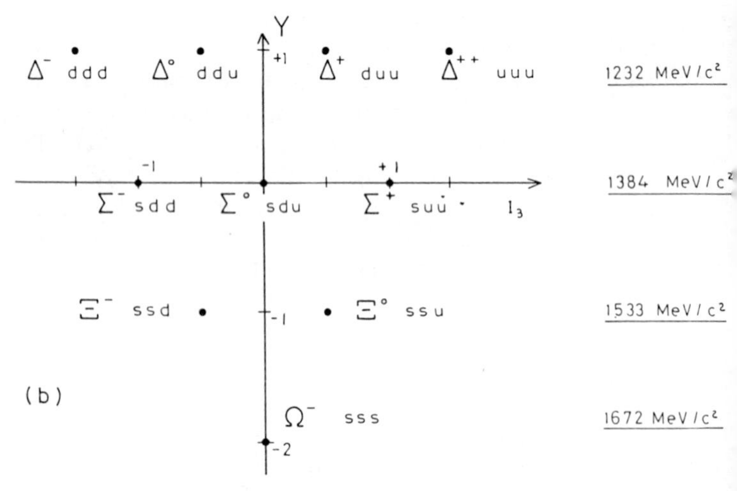

Fig. 2.3 (a) The lowest mass baryon octet and the quark content of each member. (b) The spin-3/2 baryon decuplet.

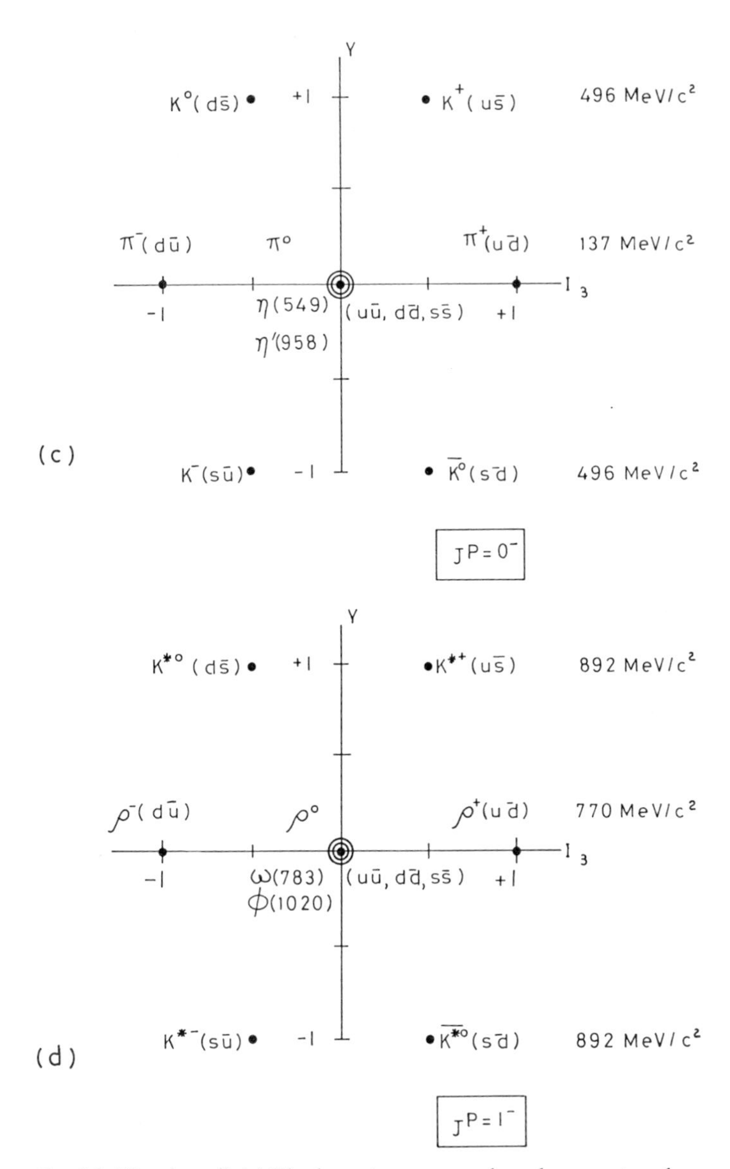

Fig. 2.3 (Continued) (c) The lowest mass pseudoscalar nonet and the quark content of each member. (d) The vector meson nonet.

At the time that the idea of quarks was mooted only nine members of the $\frac{3}{2}$ decuplet (Fig. 2.3(b)) had been observed. Gell-Mann was able to predict the existence of the final hadron in this multiplet, namely the Ω^--hyperon. Gell-Mann noted that each replacement of one of the u- or d-quarks by an s-quark leads to an increase of ~ 140 MeV/c^2 in the baryon mass; so he inferred that the Ω^--hyperon should have a mass of ~ 1680 MeV/c^2. Shortly afterwards Barnes *et al.* (1964) observed Ω^--hyperon production and decay during the exposure of a liquid-hydrogen filled bubble chamber to a beam of K^- mesons of 6 GeV/c momentum in the reaction:

$$K^- + p \rightarrow \Omega^- + K^+ + K^0.$$

The measured mass of 1672·45 MeV/c^2 for the Ω^- agrees well with the predicted value.

The weak decay of a hadron is now seen to be due to the decay of one of its component quarks. For the neutron it is a d-quark that decays:

$$\text{udd} \quad \llcorner\!\rightarrow u + e^- + \bar{\nu}_e,$$

which at the hadron level becomes

$$n \rightarrow p + e^- + \bar{\nu}_e.$$

There are, at present, five well known and one poorly established quark species or flavours; their charges and 'current' masses in MeV/c^2 are as follows:

$Q = +2/3$	u(~ 10)	c(~ 1300)	t($>40{,}000$)?
$Q = -1/3$	d(~ 10)	s(~ 150)	b(4,200).

The heavier flavours carry the names charm (c), beauty (b) and top (t). The dominant weak decays are

$$d \rightarrow u, \quad c \rightarrow s, \quad (t \rightarrow b?)$$

and these connections provide a reason for pairing the quarks as shown. It might appear that the pairing is identical with the pairing of u- and d-quarks in a strong isospin doublet.

However, this is not the case. On the one hand, the *weak* isospin pairing revealed by the dominant weak decays is connected with a gauge symmetry: on the other hand, the strong isospin symmetry is 'accidental'. More will be said about this crucial distinction in Chapters 6 and 7.

A further important source of information on the internal structure of hadrons has been the study of the scattering of high-energy leptons from nuclei. This work came to fruition in the late 1960s at SLAC with the advent of 20 GeV/c electron beams. The elastic scattering of electrons:

$$e^- + \text{proton} \rightarrow e^- + \text{proton}$$

is diffractive with a forward peak whose angular size (θ) depends on the proton size (r). If p_T is the momentum component of the scattered electron transverse to the incident beam, $p_T \approx p\theta$, where p is its total momentum. Then the uncertainty principle applied to the transverse motion gives $p_T r \sim h$ and $\theta \sim h/rp$. The scattering also shows angular variation due to the interaction itself, in which the electron and nucleus exchange a photon. Fortunately, this second dependence on angle is calculable and is multiplicative with the first. A relevant function is the Mott cross-section, which is appropriate for the elastic scattering of a point electron from a point proton. Inelastic scattering on the other hand involves the breakup of the proton in any manner:

$$e + p \rightarrow e + \text{hadrons}.$$

At SLAC Breidenbach *et al.* (1969) studied this process by measuring the energy of the outgoing electron (E') and ignoring the proton debris. Fig. 2.4 shows a comparison of the angular distributions of elastic and inelastic scattering after dividing through by the Mott cross-section so as to reveal shape effects. The variable q^2 is the four-momentum transfer, which is the appropriate relativistic invariant used to replace the angle (see Appendix A). In terms of the electron's incoming (E, $\mathbf{p}$) and outgoing (E', $\mathbf{p}'$) four-momentum:

$$q^2 = (E - E')^2 - (\mathbf{p} - \mathbf{p}')^2.$$

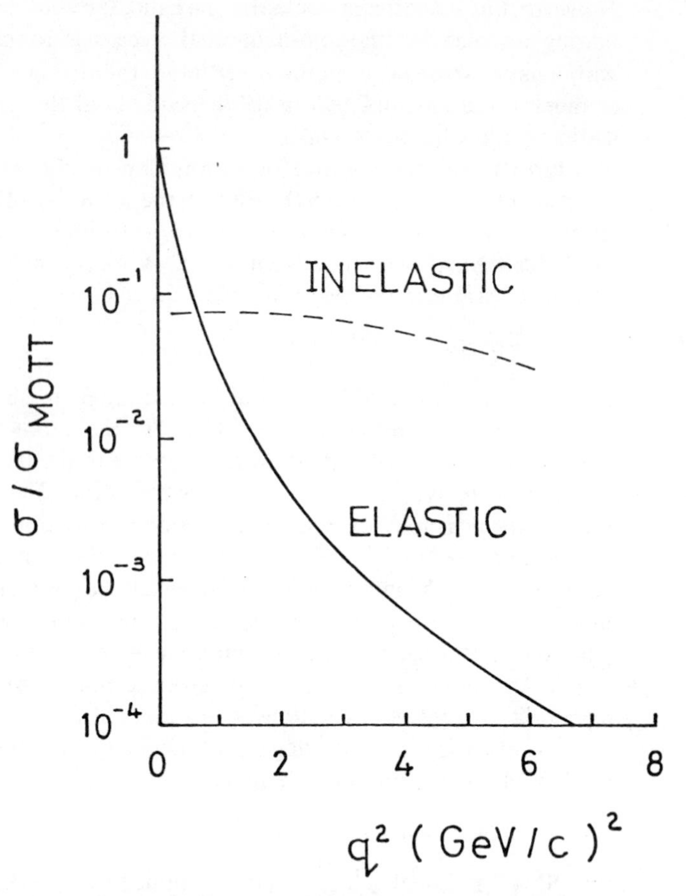

Fig. 2.4 The angular distribution of elastic and deep inelastic electron scattering from hydrogen (after Breidenbach *et al*., *Phys. Rev. Lett*. **37**, 76 (1976)).

In the present case $E \gg mc^2$, so that $E \approx pc$ and $E' \approx p'c$. If elastic scattering takes place from a massive target, then $E' \approx E$, so that the relation between four-momentum transfer and angle is quite simple:

$$-q^2 \approx 2p^2(1-\cos\theta) = 4p^2\sin^2(\theta/2).$$

Notice that q^2 is large and negative for hard collisions. There is a complete contrast in Fig. 2.4 between the shapes of the elastic and inelastic angular distributions. The inelastic distribution shows little change with angle (or q^2) and so the scattering must be taking place from something point-like. Extensions of this study with other lepton beams at Fermilab and at CERN with energies up to 400 GeV confirm this result. When a lepton of sufficient energy makes a so-called 'deep' inelastic scattering with a nucleon it appears to scatter from a substructure that is point-like. Subsequent experiments (discussed in Chapter 9) showed that these substructures are spin-$\frac{1}{2}$ fermions, and that there are (net) three per nucleon. These substructures can therefore be confidently identified with the quarks.

In conclusion we may remark that leptons and quarks behave like point-like spin-$\frac{1}{2}$ fermions. There are conservation laws for the net number of each species; quarks, electron leptons, muon leptons and tau leptons. What distinguishes the quarks is that they interact strongly. Both leptons and quarks interact weakly, and electromagnetically (if charged).

2.4 Forces

The electromagnetic force is familiar and we use it to introduce the idea of forces occurring through the exchange of *field bosons*. As an example let us consider Coulomb scattering of electrons from a nucleus which we treat as an immobile point charge Ze. The incident and outgoing electrons have four-momenta p_μ and p'_μ and can be described quantum mechanically by plane waves $\exp(-ip_\mu x_\mu)$ and $\exp(-ip'_\mu x_\mu)$ at a point $x_\mu = (t, \mathbf{r})$ in space-time. Scattering is produced by the potential

$$V(r) = -Ze^2/r$$

and the quantum mechanical amplitude for scattering in the first-order Born approximation is

$$T = \int \exp(ip'_\mu x_\mu) V(r) \exp(-ip_\nu x_\nu)\, \mathrm{d}V.$$

dV is the volume element $2\pi r^2\,dr\,d(\cos\phi)$ at the polar coordinates (r, ϕ). Then

$$T = -Ze^2 \iint \exp[i(p'_\mu - p_\mu)x_\mu] 2\pi r\,dr\,d(\cos\phi).$$

$q_\mu = p_\mu - p'_\mu$ is the change in the electron four-mementum, or four-momentum transfer. Therefore

$$(p_\mu - p'_\mu)x_\mu = q_\mu x_\mu = \mathbf{q}\cdot\mathbf{r},$$

where we ignore a small change of the electron's energy. The direction of $\mathbf{q}$ can be chosen to define the polar axis ($\phi = 0$), then

$$\mathbf{q}\cdot\mathbf{r} = qr\cos\phi,$$

$$\therefore T = -Ze^2 \int_0^\infty \int_{-1}^{+1} \exp(-iqrc) 2\pi r\,dr\,dc$$

$$= (4\pi Ze^2/q) \int_0^\infty \sin(qr)\,dr$$

$$= 4\pi Ze^2/q^2.$$

If the scattering angle is θ we have, from the previous section,

$$q^2 = -4p^2\sin^2(\theta/2)$$

so that

$$T = \pi Ze^2/p^2\sin^2(\theta/2).$$

The scattering intensity is proportional to $|T|^2$, i.e. to

$$[Ze^2/p^2\sin^2(\theta/2)]^2,$$

which recovers the well-known dependence on $\operatorname{cosec}^4(\theta/2)$. In this quantum mechanical calculation which has just been completed the amplitude is the coherent sum of contributions from the potential at all points in space. A typical individual contribution of the potential at $\mathbf{r}$ has the electron receiving the impulse at $\mathbf{r}$ exactly. This contribution can be interpreted by picturing the nucleus emitting a photon of four-momentum q which propagates to $\mathbf{r}$ and is then absorbed by the electron. The *exchange* of a photon occurs and is responsible for the force between the electron and nucleus in this view, a view which will be developed fully in chapter 4 for electromagnetism. There is an important connection between the form of the potential and the masslessness of the photon, which we now examine qualitatively.

We start from the observation that the emission or absorption of a real photon by a free particle (e.g. electron) is inconsistent with energy-momentum conservation. The equation of conservation of energy and momentum is

$(m, 0)$	$=$	$(E, \mathbf{p})$	$+$	$(E_\gamma, \mathbf{P}_\gamma)$,
parent electron		daughter electron		photon

where m is the electron mass. Momentum conservation gives

$$\mathbf{P}_\gamma = -\mathbf{p},$$

so that the photon and electron travel back-to-back. Simultaneous energy conservation cannot be achieved because

$$m \neq \sqrt{m^2 + p^2} + p.$$

However, when the photon is reabsorbed the energy-momentum balance can be restored. Such a state of affairs contradicts classical ideas but is consistent with quantum mechanics provided that the time between emission and absorption is short enough for the uncertainty principle to be satisfied. If the photon exists for only a short time, t, between emission and absorption then the precision with which the energy can be defined is subject to an uncertainty of order $\hbar/t$. Hence an energy $\hbar/t$ can be 'borrowed' for a short interval t. Photons are massless, so they may have arbitrarily small energies; t can then be arbitrarily long and the distance travelled by the photon, and hence the range of the electromagnetic force, can be infinite.

A further insight can be gained by remarking that energy and momentum conservation can be retained but only at the cost of having a photon with non-zero mass. Earlier in the section such a calculation was made to obtain the four-momentum transfer squared

$$q^2 = -4p^2 \sin^2(\theta/2).$$

This must be the four-momentum squared of the photon but it is clearly *non-zero*. An exchanged photon is said to be *virtual* because it has $q^2 \neq 0$ and only an ephemeral existence. In general a particle of rest mass m carrying a four-momentum q is said to be *virtual* if $q^2 - m^2 \neq 0$. The uncertainty relation can be used to relate the *proper* time τ for which a virtual particle can exist with an 'imbalance' $(q^2 - m^2)$ in four-momentum squared:

$$\sqrt{|q^2 - m^2|}\,\tau \sim \hbar.$$

Usually the four-momentum transfer squared involved in an exchange is negative, so that

$$\tau \sim \hbar/m,$$

which implies a finite range of force. With a particle of mass 1 GeV/c^2 the range of force (velocity $\times$ $(\hbar/m)$) is less than $c\hbar/m$, i.e. less than 0·2 f.

The field bosons which carry the weak force are extremely massive: these are the W-bosons having a mass of 83 GeV/c^2 and the Z-boson with a mass of 93 GeV/c^2. The resultant range of force is only 10^{-17} m. One consequence of this short range is that the electron-capture reaction

$$e^- + {}^A_Z X \rightarrow {}^A_{Z-1}X + \nu_e$$

only occurs when the electron overlaps the nucleus. Therefore only the capture of *K*-shell electrons is important; with increasing atomic number the degree of overlap increases and consequently the importance of this process increases.

The strong nuclear force is also of short range. Rutherford observed that the angular distribution of α-particles scattered from nuclei is consistent with the electromagnetic formula provided that the α-particles do not have enough energy to penetrate within 10^{-15} m of the nucleus. At closer approaches nuclear interactions take place and the shape of the angular distribution changes. It might reasonably be expected that the field bosons carrying the strong force are massive. This expectation led Yukawa to predict the existence of a field boson of mass ~ 100 MeV/c^2. There is indeed a plausible candidate, the π-meson, and exchange of the π-meson does play a part in hadron-hadron interactions. However, the underlying strong force between quarks involves exchange of *massless* bosons called gluons. Only the gauge theory of the strong force is capable of accounting for the limited range and other fundamental properties of the strong force. Discussion in detail is postponed until Chapter 6.

Most reactions discussed in previous sections were inelastic, e.g.

$$n \rightarrow p + e^- + \bar{\nu}_e.$$

Inelasticity is a novel feature of the fundamental forces; they can change one particle into different particles. Forces are not simply a matter of attraction or repulsion. The exchange picture is quite capable of incorporating this extended behaviour of forces. For example, if the field boson is charged the parent will change its charge on emission. We can rewrite the above weak decay by introducing the exchanged boson, W^-:

$$n \rightarrow p + W^-$$
$$\quad\quad\quad \hookrightarrow e^- + \bar{\nu}_e,$$

where the virtual W^- decays to a real electron and antineutrino. In another weak process,

$$\nu_\mu + e^- \rightarrow \mu^- + \nu_e,$$

the ν_μ emits a W^+ boson which is absorbed by the electron converting it to a neutrino.

Not only are the ranges of the three forces very different but also their strengths. The rates of processes depend on the $(\text{strength})^2$ so that decay rates are very different for the three forces. The lifetimes for two-body decays involving comparable energy release are:

weak $\mu^- \rightarrow e^- + \bar{\nu}_e + \nu_\mu$, $(Q = 105\ \text{MeV})$,
$\tau = 2{\cdot}2 \times 10^{-6}$ s,

e.m. $\pi^0 \rightarrow \gamma + \gamma$, $(Q = 125\ \text{MeV})$,
$\tau = 0{\cdot}8 \times 10^{-16}$ s,

strong $\Delta^{++} \rightarrow p + \pi^+$, $(Q = 154\ \text{MeV})$,
$\tau = 10^{-23}$ s.

The cross-sections for reactions are also very different for the three forces. At 5 GeV beam energy typical cross-sections are

weak	10^{-8} μb,
e.m.	1 μb,
strong	10 mb.

For a strong reaction an incoming beam of π^+-mesons would 'see' each proton as a disc of area 10 mb, i.e. 10^{-26} cm^2. Counting the number of π^+-mesons crossing any of these discs gives the number that scatter elastically. Cross-sections for different final states add up to give a total cross-section. Total cross-sections for strong processes on nucleons are roughly equal to the size of the proton determined from the diffraction pattern of elastically scattered electrons (10^{-26} cm^2). Thus as far as strong processes are concerned the nucleons behave like totally absorbing discs. On the other hand for the weak interaction the nucleon hardly absorbs at all: which is equivalent to full absorption over 10^{-8} μb.

Despite their differences the forces share a key common feature. They are all *gauge* forces: i.e. they are intimately connected with the symmetries of nature. This subject is explored first for electromagnetism, in Chapter 4, and for the other forces later. We shall find that the weak and electromagnetic forces are manifestations of one *electroweak* force. At the energies normally met with on earth the symmetry underlying the electroweak force is badly broken so that the weak and electromagnetic force differ greatly in strength and in range. Recognition of their common origin came only as a result of understanding the significance of gauge theories.

Chapter 3
Experimental techniques

We have seen that the μ-lepton, the positron and the π-meson were first observed as components of the cosmic radiation. Further experimental progress has relied almost exclusively on experiments using particle beams from accelerators. This is because accelerators give very intense, collimated beams of controlled energy. Experiments with such beams can be compact and can accumulate data rapidly on interesting but rare processes. For useful acceleration to be achieved a particle needs to exist for seconds, which rules out all except the stable particles; e^-, p and their antiparticles. Other particle species are produced by directing the primary beam (e.g. protons) on to a target and then sorting out the particles of interest from amongst the secondaries that emerge from the target. The quantities of interest which experimentalists wish to measure include the following. Firstly, reaction cross-sections, which contain information about the strength and space-time structure of interactions; secondly, the masses, charges and quantum numbers of particles; finally the experimentalist measures the lifetimes and decay modes of unstable particles. A wide variety of detectors has therefore been used. In the following sections accelerators and particle beams are discussed first, then detectors.

First a brief discussion of cross-sections is in order (discussed more fully in Appendix A). A cross-section for a reaction (e.g. elastic meson-proton scattering) is the equivalent area of one scatterer (e.g. proton) for this process. Suppose a hydrogen target has thickness t along the beam direction and n is the number of protons per unit volume. Then the number of protons encountered per unit area of the beam is nt. If we imagine the cross-sectional area of these protons (each of area σ) to be projected on to a surface perpendicular to the beam they cover a total area of $nt\sigma$. Any meson incident on this area is scattered; any meson incident on the remaining area $(1-nt\sigma)$ (≈ 1) is not scattered. Then if there are N mesons incident the number scattered is $N_s = N(nt\sigma)$.

In order to measure a cross-section it is enough to measure N and N_s and to know the target parameters n and t. Hydrogen and deuterium targets are preferred because the nuclei are simple. A deuterium target consists of virtually free protons and neutrons because the deuteron binding energy is so small. Measurements on complex nuclei are harder to interpret because multiple collisions can occur inside nuclei and because the residual nucleus after a collision can undergo spallation. The hydrogen or deuterium is used in liquid form so that the target can be compact. When measuring angular distributions of outgoing particles from a reaction, a set of detectors can be arranged around the target to cover the solid angle of interest. Suppose $\Delta\Omega$ is the solid angle subtended at the target by one element of the detector. Then the number of mesons scattered into this element is

$$N_s = Nnt(\mathrm{d}\sigma/\mathrm{d}\Omega)\Delta\Omega.$$

$\mathrm{d}\sigma/\mathrm{d}\Omega$ (measured in $\mathrm{cm}^2\ \mathrm{sterad}^{-1}$) is called a *differential cross-section* and refers to a particular direction of scattering. It will generally depend on the scattering angle. Integration over all directions yields the total cross-section for the reaction studied

$$\sigma = \int(\mathrm{d}\sigma/\mathrm{d}\Omega)\ \mathrm{d}\Omega.$$

The elastic cross-section (when only the beam and target particles emerge from the interaction) is written σ_{EL}. Other, inelastic, reactions are lumped together to give σ_{R}, the reaction cross-section. The total cross-section σ_{T} is $(\sigma_{\mathrm{EL}} + \sigma_{\mathrm{R}})$.

3.1 Accelerators and beams

Almost all accelerators used in particle physics employ cyclic acceleration. The cyclotron, which was the first of this type, consists of a flat metal cylinder placed between the flat poles of an electromagnet (Fig. 3.1). This metal can is sliced along a diameter into two separate 'Dees' and at the centre is placed a hydrogen-ion (proton) source. Protons circulating inside the Dees in a plane perpendicular to the field of the electromagnet (**B**) feel a Lorentz force evB perpendicular to **B**. The force is also perpendicular to **v**, so that no work is done (classically); the proton is steered but $|\mathbf{v}|$ is not changed. If the radius of curvature of path under this lateral force is instantaneously r, then

$$m\gamma v^2/r = evB.$$

Therefore the proton follows a circular path with angular frequency

$$\omega_c = v/r = eB/m\gamma,$$

where ω_c is the cyclotron frequency, $\gamma = 1/\sqrt{(1-\beta^2)}$ and $v = \beta c$. Provided velocities are low compared to c then ω_c does not depend on v or r. A voltage applied between the Dees causes the protons circulating in one sense to accelerate on crossing the gap; inside the Dees they coast on a circular path. In the cyclotron the applied voltage alternates at angular frequency ω_c so that if a proton is accelerated at one crossing it is accelerated at every crossing. Between crossings such protons coast on orbits of increasing radii. When the accelerated protons reach the outside edge of the Dees they can be electrostatically deflected out for use in experiments.

There are difficulties, however. At relativistic energies the mass of the proton increases to the extent that the cyclotron frequency alters and continuous acceleration is no longer possible. Another difficulty of the system is that a huge magnet would be needed for acceleration to GeV energies. Modern synchrotrons accelerate 'bunches' of protons from a low energy up to the top energy on a fixed orbit so that the magnets can be of manageable size. In order to maintain a fixed orbit it is

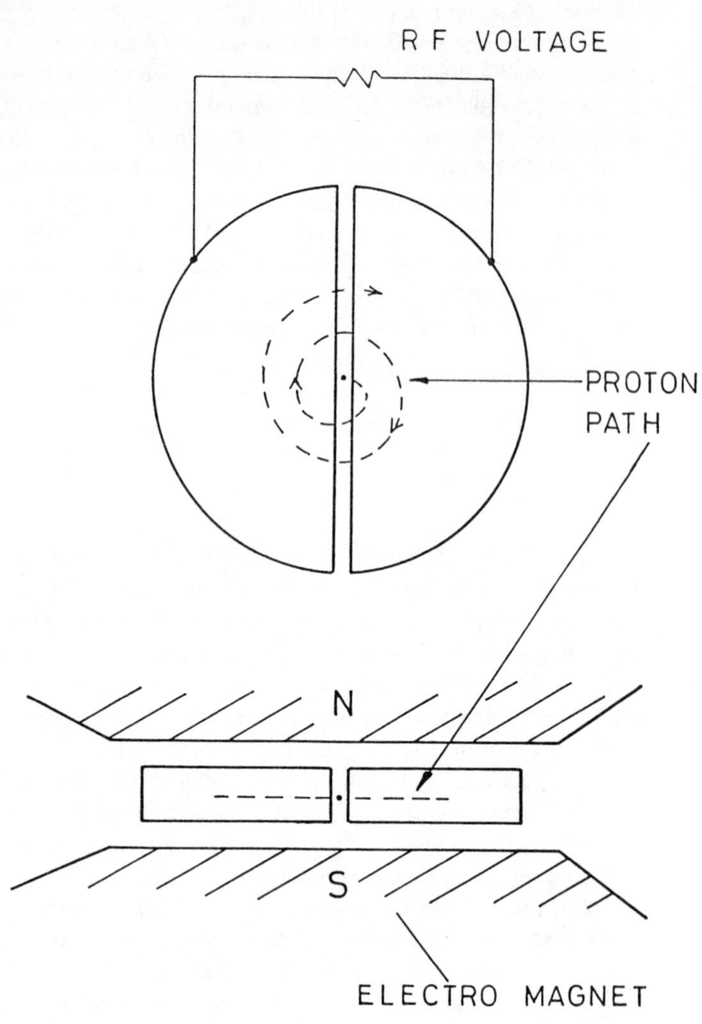

Fig. 3.1 Vertical and horizontal sections through a cyclotron.

necessary for the magnetic field to increase by a large factor through the cycle. Acceleration is provided by RF cavities placed around the ring and the RF frequency also increases slightly during an acceleration cycle. Once acceleration has been achieved the beams can coast round or be ejected for use in experiments.

The three essential elements of the cyclic accelerators are: RF cavities to provide acceleration, dipole magnets to steer the protons around the ring and quadrupole magnets to focus the beam. Fig. 3.2 shows the pattern of magnetic field lines in a section through a quadrupole magnet taken perpendicular to the beam axis. A proton moving into the diagram is focused at A or A′ but defocused at B or B′. The opposite effect of vertical

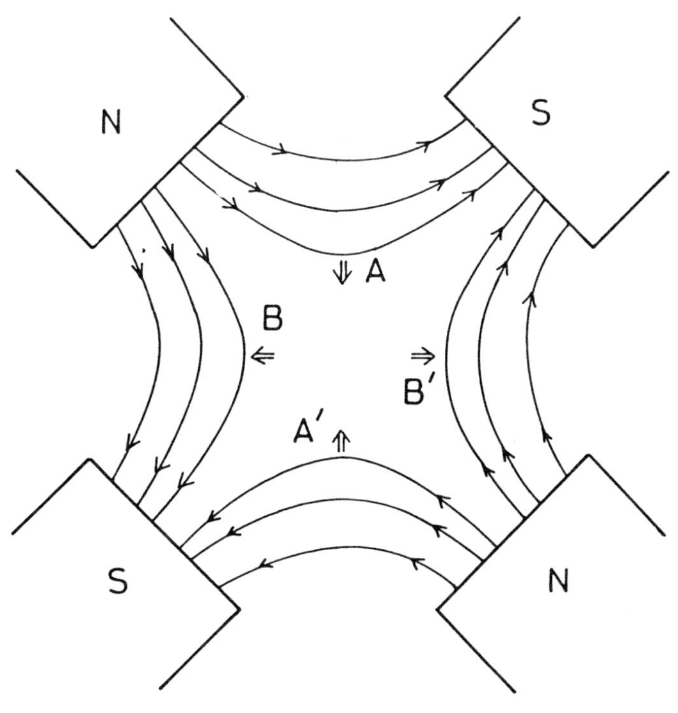

Fig. 3.2 The field pattern of a magnetic quadrupole.

defocusing and horizontal focusing is achieved by reversing the polarity of the coils. As with optical elements a net focusing, in both planes, can be achieved with alternately arranged vertical and horizontal focusing magnets.

At CERN the protons are first accelerated to 25 MeV in a linear accelerator, injected into the proton synchrotron (PS) of radius 100 m and accelerated to 25 GeV in about 1 second. The field in the bending magnets rises from 15 mT to 1·4 T during this period. Subsequently bunches of protons can be injected into the super proton synchrotron (SPS) of radius 1·1 km and accelerated by two 200 MHz RF cavities to an energy of 400 GeV in about 3·7 seconds; the protons gain 2·54 MeV energy per turn. The RF imposes a structure on the beam so that it consists of 4 620 bunches. Each circuit of the protons takes only 23 μs. Powerful kicker magnets and electrostatic elements are used to eject the protons for use.

Synchrotrons are also used to accelerate electrons. In the case of electrons the synchrotron radiation which all charged particles emit when deviated by a magnetic field becomes important. The energy loss per turn for a particle whose velocity is close to c is

$$\Delta E = (4\pi/3)\,(r_e/r)\gamma^4(m_e c^2)$$

when following a path of radius r. Here r_e is the classical electron radius, 2·82 f. For electrons the energy loss is 10^{13} times that for protons at the same energy in the same ring. In the electron-positron storage ring at DESY with a beam energy of 20 GeV and an arc radius of 256 m the loss per turn is 64 MeV. Considerable RF power (8 MW) is needed to maintain the beam energy. At CERN the LEP electron-positron accelerator now under construction will have an energy per beam of 50 GeV. A radius of 4·3 km is needed in order to restrain its power requirements.

The highest centre-of-mass (CM) energies are obtained from accelerators in which particles from two counter-rotating beams collide. (The current energy record of 2 TeV is held by the $\bar{\text{p}}$p collider at Fermi National Accelerator Laboratory (FNAL) near Chicago, Illinois.) It is shown in Appendix A that the CM energy in a collision between a beam proton of mass m and energy E and a target proton at rest is

$$W=\sqrt{2mE} \text{ if } E \gg m.$$

However, if an antiproton and proton, both of energy E, collide head on the CM energy is now

$$W=2E.$$

A complex sequence of operations needs to be followed to generate, store and accelerate the antiprotons. Successful operation was first achieved at CERN and is illustrated in Fig. 3.3. Antiprotons are produced by firing 28 GeV protons

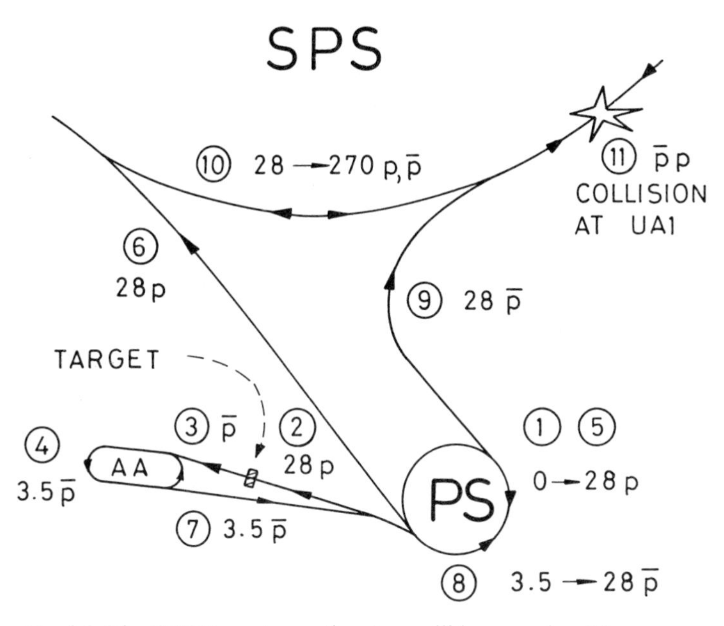

Fig. 3.3 The CERN proton-antiproton collider complex. The numbers within circles indicate the chronological order of steps in the sequence to obtain colliding beams inside the SPS. The beam energies in GeV at each stage are also indicated (from *Eur. J. Phys.* **6**, 41 (1985)).

from the PS on to a fixed target. The yield is low, only one $\bar{p}$ per 10^6 protons. These antiprotons are steered into a storage ring (AA) where they accumulate over a day. At entry they have a wide range of momenta such that if injected directly into a synchrotron most would be lost. The relative motion between the antiprotons (the thermal motion) needs to be 'cooled' so that they form a useful beam. Van der Meer devised an elegant way to cool the beam of antiprotons in the accumulator. Measurements are made of the location of the charge by pick-up electrodes at one point on the orbit and a correction voltage applied via electrodes at the diametrically opposite point. It is just possible to transmit the signal across the diameter quicker than the antiprotons can travel round the arc. The signals lead stochastically to a reduction of the relative momenta. When $\sim 10^{11}$ antiprotons have been cooled and collected the loading phase commences. Protons are accelerated in the PS to 28 GeV and transferred to the SPS. Next the accumulated antiprotons are transferred in bunches from the AA to the PS where they, too, are accelerated to 28 GeV and then transferred to the SPS. At this moment there are counter-rotating bunches of protons and antiprotons inside the same beam pipe in the SPS, each at 28 GeV energy. The beams can now be simultaneously accelerated because whatever arrangement of electric fields accelerates protons in one sense will automatically accelerate the antiparticles (antiprotons) in the opposite direction. The beams are simultaneously accelerated to anywhere between 100 and 310 GeV and then continue to circulate at the chosen energy. The protons and antiproton bunches are brought into collision at selected locations around the SPS ring at places where there are underground experimental areas (UA). Beam losses are small, so that it is only necessary to refill the collider once per day. The interaction rate can be calculated from the expression

$$N = [n(\mathrm{p})n(\bar{\mathrm{p}})f/a]\sigma \text{ sec}^{-1},$$

where $n(\mathrm{p})$ and $n(\bar{\mathrm{p}})$ are respectively the number of protons in one proton bunch and antiprotons in one antiproton bunch, f is the frequency of rotation and a is the bunch cross-sectional area. These areas are assumed to overlap exactly on crossing. The factor in the square brackets is called the luminosity ($cm^{-2}\ sec^{-1}$). With three bunches of protons and of antiprotons in the machine the luminosity (L) is increased by factor 3. At CERN luminosities of $3 \times 10^{29}\ cm^{-2}\ sec^{-1}$ have been achieved. The cross-section for non-diffractive processes at 540 GeV CM energy is ~ 55 mb ($55 \times 10^{-27}\ cm^2$) so that the interaction rate at a crossing point is $\sim 16\,500$ interactions per second.

Unstable particles can be generated by directing the primary proton beam on to a target, which might be a tungsten block. All manner of hadrons emerge in all directions and with a range of momenta. Focusing magnets are used to collect particles emerging in as large a solid angle as possible. Momentum selection is made using collimating slits in metal plates and bending magnets. This is illustrated in Fig. 3.4 where a beam of particles with a range of momenta is incident from the left and is focused at the plane of the central slit (collimator). The trajectories of particles at the design momenta are shown as solid lines. Broken lines are used for the trajectories of lower momenta particles which turn more in the magnetic fields. The size of the collimator aperture determines the range of momenta transmitted. A second bending magnet acts to compensate the

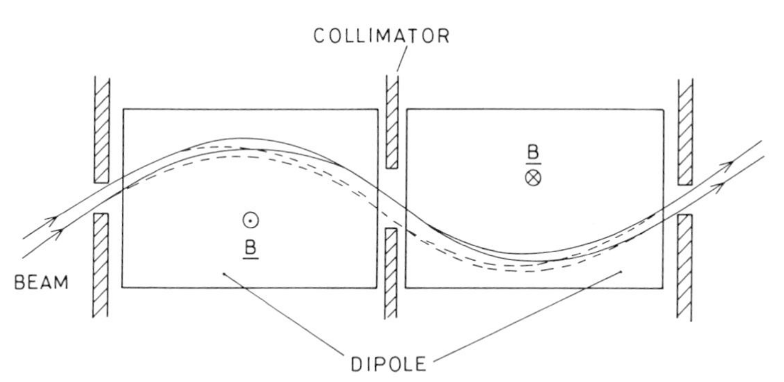

Fig. 3.4 A simple arrangement for momentum selection with a second dipole magnet compensating the dispersion introduced by the first.

dispersion produced by the first; thus all particles passing through the collimator converge into a single beam. (An optical analogy is helpful here: dipole magnets are equivalent to prisms and quadrupoles are equivalent to lenses.) What then remains is to select the particle type. At low energies (a few GeV) electrostatic selection is possible making use of the difference in deflection in an electric field of particles having equal momenta but different masses. Enrichment of one particle species is feasible up to 40 GeV/c using RF cavities. Alternatively no attempt at separation may be made and instead the particles in the beam are labelled using signals from Cerenkov counters which they traverse. The operation of Cerenkov counters relies on the fact that a charged particle travelling at a speed greater than the local speed of light in a material will radiate Cerenkov radiation. This radiation is analogous to the shock wave from an aircraft when it exceeds the speed of sound. The Cerenkov radiation emerges as a conical wavefront travelling at an angle θ with respect to the beam axis, such that

$$\cos\theta = 1/\beta n,$$

where n is the refractive index of the medium and βc is the particle velocity. Therefore in a beam of particles, all at the same *momentum*, the π-mesons will emit Cerenkov radiation at some angle and the K-mesons at a different angle. If the active material of the Cerenkov counter is chosen so that

$$\beta_{\mathrm{K}} < 1/n \text{ but } \beta_{\pi} > 1/n$$

it follows that *only* π-mesons will emit Cerenkov radiation. This is the threshold mode of operation. Photomultipliers are used to view the active material and detect the presence or absence of Cerenkov radiation.

Once the selection or identification of the particle type and the momentum selection have been made the secondary beam is steered and focused on to the experimental apparatus. During its passage it is important that the beam passes through as little material as possible because scattering in matter changes the momentum whilst interactions also generate unwanted particles. Where possible an evacuated beam pipe is employed.

Neutrino beams from accelerators require a special approach. Mention was made in Chapter 2 of a neutrino beam involving no momentum selection, a so-called 'wide-band' beam. Fig. 3.5 shows the components of the '*narrow*-band' beam line at CERN, which gives a pure neutrino or antineutrino beam with some momentum selection. The primary proton beam (10^{13} protons) is dumped every machine cycle (~ 6 s) on to a target in the form of a metre-long beryllium rod. Neutrinos originate from the two-body decays of π- and K-mesons produced in the target:

$$\pi^+ \rightarrow \mu^+ + \nu_\mu \quad \text{(branching fraction} \sim 100\%\text{)},$$
$$K^+ \rightarrow \mu^+ + \nu_\mu \quad \text{(branching fraction} \sim 63{\cdot}5\%\text{)}.$$

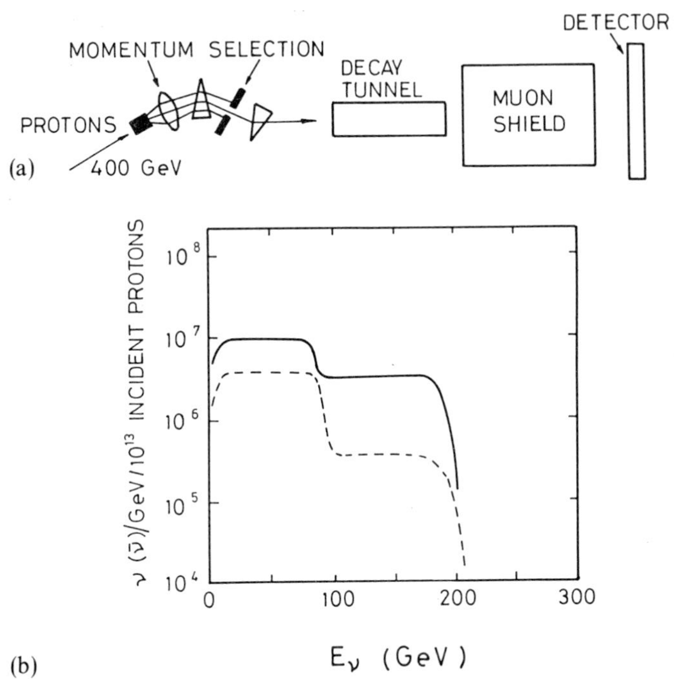

Fig. 3.5 (a) The components of the neutrino narrow-band beam line at CERN. (b) The energy spectrum of the neutrinos (solid curve) and antineutrinos (broken curve) (After Pullia, *Riv. Nuovo Cim.* **7**, 2 (1984)).

The light hadrons rarely decay with the emission of electron neutrinos, so that there are few electron neutrinos in the beam. Quadrupoles placed immediately after the target (drawn as a lens in Fig. 3.5) offer a large angular acceptance. Then magnetic dipoles (drawn as prisms) are used to select particles of one sign and of momenta in the range 200 ± 10 GeV/c. With positive (negative) sign selected the beam will be pure neutrino (antineutrino). After sign and momentum selection the beam traverses an evacuated tube 300 m long where a percentage of the mesons decay giving neutrinos. Beyond the tubes lies a 400 m thick muon shield of iron and earth where the remaining hadrons and charged leptons are absorbed and brought to rest. Bringing the μ^+-leptons to rest before they decay is crucial: the energy of their daughter antineutrinos is then only the decay energy and their presence does not compromise the purity of the high-energy neutrino beam. Fig. 3.5 shows both the neutrino spectrum (solid line), and the alternative antineutrino spectrum (broken line) obtained if negative hadrons are selected. The two-body decays of π- and K-mesons are isotropic in the parent rest frame so that boosting to the laboratory frame gives a flat spectrum in each case. The maximum laboratory energy of the neutrino is given by:

$$(\pi \text{ parent})\ E_{\max} = E(1 - m_\mu^2/m_\pi^2) = 0{\cdot}43E$$
$$(\text{K parent})\ E_{\max} = E(1 - m_\mu^2/m_\text{K}^2) = 0{\cdot}95E,$$

where E is the parent energy. When the two spectra are added the result is the two-step spectrum seen in Fig. 3.5. There will also be a correlation between the decay angle of the neutrino and its energy. For example, a neutrino emitted exactly forward will have the maximum energy. It is therefore possible to infer the neutrino energy to an accuracy of 10 per cent from the lateral distance between its interaction point and the beam axis.

Monitoring methods for neutrino beams are necessarily indirect. One method is to measure the muon flux in the muon shield and another is to measure the total current in the hadron beam.

Differential Cerenkov counters have also been used to

measure the K- and π-meson beam fractions. The narrow band beam has a 3 per cent component of the 'wrong' energy neutrinos from three-body decays $K \to \pi + \mu + \nu_\mu$ and less than 0·1 per cent contamination of antineutrinos in the neutrino beam. Neutrinos from the decay of hadrons near the target and from hadrons not accepted by the momentum selection could be a serious background. For this reason the primary proton beam is turned through 15 mrad before it strikes the target, and any such backgrounds are directed away from the detector. Wide-band beams have about a hundred times greater flux than narrow-band beams. They suffer from high backgrounds, are less easy to monitor and lack any angle-energy correlation.

3.2 Particle interactions with matter

Charged particles transfer energy to the medium through which they travel by ionization and this process forms the basis of particle detection. Neutral particles can only be detected if they first transfer energy to charged particles. The dominant process for transfer of energy from photons (above a few MeV) is for them to convert to an electron-positron pair in the Coulomb field of a nucleus. Neutral hadrons can decay to charged particles; for example, a K^0-meson can decay to a π^+- plus a π^--meson. π^0-mesons require a two-step process: first they decay to two photons, which then convert to electron-positron pairs. Finally neutrons, which have relatively long lives, can undergo nuclear interactions; the most efficient in terms of energy transfer being elastic scattering from protons.

The ionization loss per unit path length of a singly charged particle with velocity βc is given by the Bethe-Bloch formula:

$$dE/dx = (D\rho Z/A\beta^2)[\ln(2m_e c^2\beta^2\gamma^2/I) - \beta^2] \text{ MeV m}^{-1}$$

where

$$D = 4\pi N_0 r_e^2 m_e c^2 = 0{\cdot}0307 \text{ MeV m}^2 \text{ kg}^{-1}.$$

N_0 is Avogadro's number, r_e is the classical electron radius and m_e its mass; Z is the atomic number for the medium, A its atomic weight and ρ its density in kg m^{-3}. I is the effective ionization of the medium. At low velocities the ionization loss is proportional to $1/\beta^2$; the ionization loss reaches a minimum at $\gamma \sim 3$ and then slowly rises toward a plateau at very large values of γ. The extent of this rise depends on the medium, and is of order of 25 per cent. Fig. 3.6 shows experimental measurements made of ionization loss in a gas-filled drift chamber (Kubota *et al.* (1983)). At a fixed momentum the velocities of protons, K-mesons, π-mesons and electrons differ, whence their ionization losses differ. The observed spread of measured values in Fig. 3.6 is due to statistical and other sampling errors. Energy losses at the ionization minimum are 10·7 MeV cm^{-1} in iron, 0·2 MeV cm^{-1} in liquid hydrogen and 2·4 keV cm^{-1} in gaseous argon. In this last case the effective ionization potential is 16 eV, of order ninety ion pairs are formed per cm of path. Ionization loss is small and detection must rely on some form of

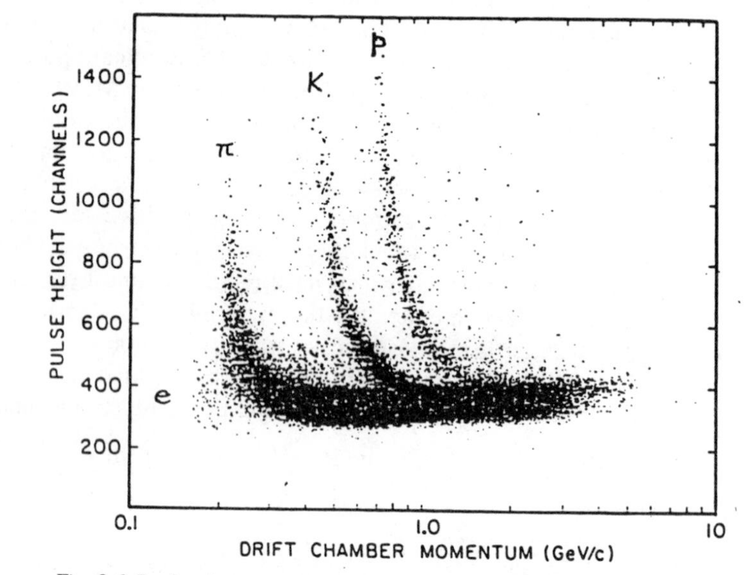

Fig. 3.6 Ionization measurements made in a drift chamber (after Kubota *et al.*, *Nucl. Inst. Meth.* **217**, 249 (1983)). The measurements terminate at 200 MeV so that the rise for electrons at low momentum is not observed.

amplification to get macroscopic signals. In certain detectors the medium is in a meta-stable state from which the ionization triggers a change of state. Nuclear emulsions and bubble chambers are in this category. A variety of other detectors rely on electrical amplification. Before discussing detectors in detail, however, it will be useful to describe other mechanisms by which particles interact with matter.

Electrons, which are relatively light, can radiate an appreciable part of their energy as photons when decelerated in the intense electric field close to a nucleus. This radiation is called bremsstrahlung. The mean energy loss by bremsstrahlung per unit path is

$$\mathrm{d}E/\mathrm{d}x = E/X_0,$$

where X_0 is the 'radiation length' of the material: in one radiation length the energy of an electron is reduced on average by a factor $1/e$. At low energies ionization loss dominates; at high energies the bremsstrahlung loss dominates. The energy at which the losses become equal is called the critical energy, E_c, given approximately by

$$E_c = 0{\cdot}6\ \mathrm{GeV}/Z,$$

where Z is the atomic number of the medium. A bremsstrahlung (or any) photon with energy above $2m_e c^2$ can convert to an electron-positron pair in the electric field of a nucleus. The mean probability for conversion per unit path is (above ~ 1 GeV energy)

$$\mathrm{d}P/\mathrm{d}x = 7/(9X_0).$$

Both bremsstrahlung and conversion are electromagnetic processes occurring in the nuclear Coulomb field, so the rates are proportional to the atomic number of the medium squared (Z^2). X_0 is therefore short in high-Z materials like lead (0·56 cm) but long in liquid hydrogen (~ 10 m).

Electrons and positrons which emerge from a conversion can, if energetic enough, radiate further bremsstrahlung photons which can in their turn convert. An electromagnetic shower develops containing a roughly equal mix of electrons, positrons and photons. Multiplication proceeds until the mean particle energy falls below the critical energy and subsequently the shower dies away. Electromagnetic showers will develop if the primary particle is a photon, electron, positron or a neutral hadron (like the π^0-meson), which decays to photons. The radiation length determines the distances over which a shower develops and within which it is contained. A mathematical analysis of shower development reveals a result crucial for making energy measurements on photons, electrons, etc.: that the total track length of all electrons and positrons in the shower is proportional to the energy of the parent. It is fair approximation to assume that in one radiation length a photon of energy E converts to an electron and a positron, each of energy $E/2$; and that an electron or positron loses half its energy in one radiation length. After n radiation lengths the current number of photons plus electrons plus positrons is 2^n and each has an energy $E/2^n$. Multiplication continues until $E/2^n$ falls below the critical energy E_c. This happens after a number of generations

$$n_c = \ln(E/E_c)/\ln 2.$$

At this stage the shower contains a maximum number of particles

$$N_c = E/E_c.$$

Suppose the remaining electrons and positrons travel a further mean distance r before absorption. Thus the total track length of positrons plus electrons summed over each generation of the shower is

$$L = (2/3)\left\{N_c r + \sum_{p=1}^{n_c} 2^p X_0\right\}$$

$$= (2/3)(r + 2X_0)(E/E_c),$$

showing that the total length is proportional to the parent energy. The shower peak occurs at a depth which increases with energy. In lead, for which E_c is 6·9 MeV, the peak of a 50 GeV shower occurs after

$$n_c = \ln(50/0{\cdot}0069)/\ln 2$$
$$= 12 \text{ radiation lengths.}$$

The energy of such a shower is contained in some 20 radiation lengths or about 12 cm of lead.

The light hadrons are two to three orders of magnitude heavier than electrons and have negligible energy losses due to bremsstrahlung at currently available energies. However, hadrons interact strongly with nuclei when passing through matter and produce more hadrons. These secondary hadrons can in turn interact so that a *hadron* shower develops. The characteristic length is in this case the interaction length (λ) for the medium; after travelling a distance λ the fraction of hadrons that have *not* undergone an inelastic interaction is $1/e$. The interaction length is 700 cm in liquid hydrogen but only 17 cm in iron or lead. A measure of the penetration of a hadron shower is the depth needed to contain 95 per cent of the incident energy; in iron it is about 80 cm for a hadron of 50 GeV. This gives an indication of the size and weight of practical calorimeters.

Muons behave like electrons in matter but are 200 times more massive. Therefore at accessible energies they ionize but do not give appreciable bremsstrahlung. Thus muons are identified by their characteristic single penetrating tracks at high energy. There is no hadronic or electromagnetic shower. Finally, neutrinos interact only weakly and generally escape detection. Their presence may be inferred if all other products from an interaction are detected yet an energy imbalance remains.

3.3 Bubble chambers

Two examples of visual detectors have been mentioned in Chapter 2, namely cloud chambers and nuclear emulsions. A third type of visual detector is the bubble chamber. This consists of a vessel of liquid held at a temperature near its boiling point. It is placed where an external beam from an accelerator will pass through its volume. The liquid is expanded in synchronism with the accelerator cycle so that, when the beam from the accelerator traverses it, the liquid is in a superheated condition. Each charged particle leaves ions along its path; these give rise to local hot spots on which bubbles can form in the superheated liquid. A few milliseconds later a set of simultaneous stereo-photographs are taken using flashes. At this stage the bubbles have grown to a size of 5–50 μm and can be seen as bright spots with dark-field illumination, i.e. the images of the flash tubes are focused away from the cameras, which then only receive light scattered by the bubbles. Once the photographs have been taken the liquid is recompressed to collapse the bubbles and to await the next beam pulse.

In bubble chambers filled with liquid hydrogen the liquid acts as a simple target (protons) and forms a detector at the same time. A common feature is to have a magnetic field (B teslas) parallel to the optic axes of the cameras. Then the particle momentum (p GeV/c) can be deduced from the track curvature (ρ metres):

$$p = 0{\cdot}3B\rho$$

when the particle carries charge $^{\pm}e$. For such a particle the sagitta of its track over a length L metres is (see Exercise Q3.2 in Appendix F):

$$s = 0{\cdot}3BL^2/8p.$$

If the mid point and end points of the track are measured on the stereographs the value of p which is obtained will have an error Δp_{m},

$$\Delta p_{\mathrm{m}}/p = \sigma_x/s = 8\sigma_x p/0{\cdot}3BL^2,$$

where σ_x is the error in the position measurement. This error can be reduced by making N equally spaced measurements along the track. It becomes (Gluckstern (1963)):

$$\Delta p_{\mathrm{m}}/p = (\sigma_x p/0{\cdot}3BL^2)\sqrt{720/(N+4)}.$$

Thus if 50 points are measured with 100 μm precision along a metre of track recorded in a 1 T field,

$$\Delta p_{\mathrm{m}} = 0{\cdot}0012p^2.$$

A second important contribution to the error in momentum measurement comes from multiple scattering, i.e. the many uncorrelated scatters that a charged particle undergoes in collisions with atoms. This error has a magnitude given by

$$\Delta p_{ms}/p = (0{\cdot}06/BL)\sqrt{L/X_0}.$$

In the case just considered, when the medium is liquid hydrogen with a radiation length of 10 m,

$$\Delta p_{ms} = 0{\cdot}02p.$$

At low momenta and for materials of high atomic number the multiple-scattering error dominates, while at high momenta the measurement-error dominates.

The spatial resolution of bubble chambers is well matched to the study of decays of strange and charmed hadrons. In a mean lifetime, τ, a particle of velocity βc travels a time-dilated distance

$$l = \tau\beta\gamma c = c\tau(p/m),$$

where $\gamma = (1-\beta^2)^{-\frac{1}{2}}$ and p is again the momentum. A $\Lambda^0(1116)$ of momentum 2 GeV/c will travel a distance 14·2 cm in its mean lifetime of $2{\cdot}63 \times 10^{-10}$ s and a $D^0(1865)$ of the same momentum will travel 140 μm in its mean lifetime of $4{\cdot}4 \times 10^{-13}$ s. Small bubble sizes and precision optics are necessary to facilitate a lifetime measurement in the second case.

The measurements of the $\Lambda^0(1116)$ mass and lifetime have been made using reactions such as

$$K^- + p \rightarrow \Lambda^0 + \pi^0$$
$$\hookrightarrow \pi^- + p.$$

In a hydrogen bubble chamber of dimension ~ 2 m the production and decay may both be seen if the beam momentum is a few GeV/c. The track curvatures determine the momenta of the π^--meson and proton ($\mathbf{p}_\pi$ and $\mathbf{p}_p$); and hence their energies (E_π and E_p) using the relation $E = \sqrt{p^2 + m^2}$. These values fix the $\Lambda^0(1116)$ four-momentum: from energy-momentum conservation

$$E_\Lambda = E_p + E_\pi \quad \text{and} \quad \mathbf{p}_\Lambda = \mathbf{p}_p + \mathbf{p}_\pi,$$

whence the mass is determined using

$$m_\Lambda = (E_\Lambda^2 - p_\Lambda^2)^{\frac{1}{2}}.$$

The current best value is $1115{\cdot}60 \pm 0{\cdot}05$ MeV/c^2. Each event also gives a lifetime for one $\Lambda^0(1116)$. Its path length (l) is the distance from the point at which the track of the incident K^--meson disappears to the point from which the π^--meson and proton tracks emerge. Then the lifetime of this $\Lambda^0(1116)$ is

$$t = m_\Lambda l / p_\Lambda c.$$

When a plot of the frequency distribution for survival longer than a time t is plotted against t for many such decays, an exponential shape emerges:

$$f(t) = f(\mathrm{o})\exp(-t/\tau_\Lambda).$$

A fit to the slope determines τ_Λ, whose current experimental value is $2{\cdot}632 \pm 0{\cdot}020\ 10^{-10}$ s.

Fig. 3.7 illustrates the single-event capability of the bubble-chamber technique. This displays the first example of the observation of Ω-production and decay: the line diagram to the right shows the identity of each particle in the reaction chain:

$$K^- + p \rightarrow \Omega^- + K^+ + K^0$$

$\quad\quad \hookrightarrow \Xi^0 + \pi^-$ (weak decays)

$\quad\quad\quad \hookrightarrow \pi^0 + \Lambda^0$ (weak decays)

$\quad\quad\quad\quad \Lambda^0 \hookrightarrow \pi^- + p$ (weak decays)

$\quad\quad\quad \pi^0 \hookrightarrow \gamma + \gamma$

$\quad\quad\quad\quad \hookrightarrow e^+ + e^-$ (pair production)

$\quad\quad\quad\quad \hookrightarrow e^+ + e^-$ (pair production)

Track curvatures were measured in order to determine the momenta of decay products, e.g. the π^- and p from Λ^0-decay. In this case the vector sum of these momenta is the Λ^0 momentum: which enabled the experimenters to draw a line back from the decay point to indicate the flight path of the neutral Λ^0. The

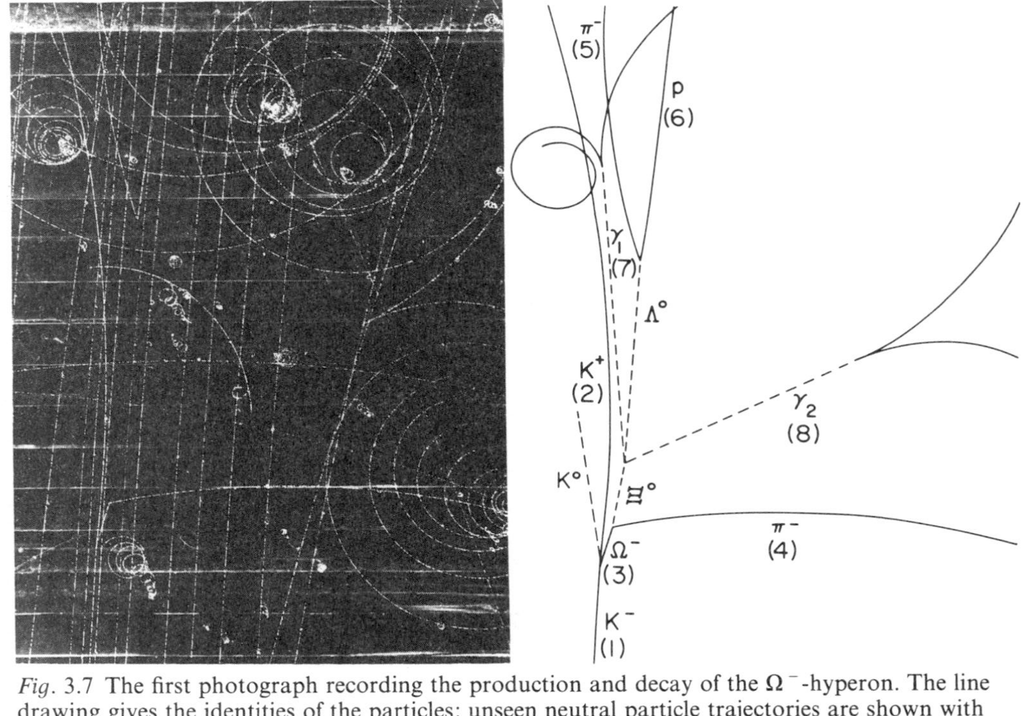

Fig. 3.7 The first photograph recording the production and decay of the $\boldsymbol{\Omega}^-$-hyperon. The line drawing gives the identities of the particles: unseen neutral particle trajectories are shown with broken lines (from V. Barnes *et al.*, *Phys. Rev. Lett.* **12**, 204 (1964)). Photo courtesy Brookhaven National Laboratory.

observed convergence of the Λ^0, γ_1 and γ_2 flight paths to a point indicates their common origin. Similar kinematic arguments fixed the whole sequence and this one event established the existence of the Ω^--hyperon.

Bubble chambers have been excellent devices for measuring lifetimes, for establishing common decay modes and for measuring the masses of excited hadrons (see Chapter 10). However, they are of less value for the study of rare processes and for accumulating high statistics. Their operation is slow, and data reduction from film is slow; finally they cannot be triggered to record only those events of interest because they must expand before the arrival of the beam. They have been superseded in great measure by fast devices which are triggerable or which are virtually continuously sensitive and whose electronic readout is triggerable. Wide-gap drift chambers, which belong to the second category, are discussed in the next section.

3.4 Wire chambers

These chambers contain one or more thin anode wires and a negative electrode, which may be an array of wires or a foil all immersed in a suitable gas. When charged particles pass through the gas they leave ion pairs. The electrons drift to the anode wire where the field can be made large by having a small wire radius (25–100 μm). The field E at a distance r from a wire radius r_i located inside a cylinder radius r_0 at a relative potential V is

$$E = V/[r \ln(r_0/r_i)].$$

If the field near the wire is sufficiently strong the electrons acquire enough energy between collisions to produce further ion pairs. An amplification of the charge deposited at the wire compared to the primary ionization of 10^4 to 10^5 is possible, with the pulse collected being *proportional* to the number of primary ions. Multiwire proportional chambers (MWPCs) consist of a plane of anode wires between parallel foil cathodes. Typical dimensions are, 25 μm diameter anode wires with

2·5 mm spacing held at ground potential and cathode planes at 5 mm distance held at −5 KV. The signal on an individual wire due to the avalanche around it is not shared with other wires; the electrons travelling toward the anode wire are shielded by a sheath of positive ions moving away from the wire (Charpak (1968)). The property means that such devices can be used for position measurement. Particle positions can be recorded to an accuracy better than 1 mm and the pulse defines the timing to better than 50 ns. Very large area (10 m^2) MWPCs have been widely used. Spatial resolution is relatively poor compared to the bubble chamber but a very high interaction rate can be tolerated because the MWPC voltage levels recover rapidly after a pulse.

A more sophisticated wire detector is the wide-gap drift chamber. An example is that used in the UA1 experiment at CERN to study the products of proton-antiproton collisions. It forms a cylindrical shell surrounding the beam pipe; it has an inner diameter of 0·2 m, and an outer diameter of 2·4 m, and a length of 6 m. Sections through one of the six semi-cylindrical elements of the drift chamber are shown in Fig. 3.8. The basic cell consists of a plane of cathode wires at −30 KV and a parallel anode plane at a distance of 18 cm. The anode plane wires are spaced at 5 mm separation with alternately a sense wire (diameter 35 μm) and a field wire (diameter 100 μm). Not shown in the figure is a third layer of field wires between the two anode planes. Around the cover conductive strips and a resistive chain ensure a uniform potential drop over the anode-cathode gap. The gas in the detector is 40 per cent argon and 60 per cent ethane at NTP.

When a charged particle passes through the gas it leaves ion pairs. The electrons drift at about 50 mm μs^{-1} toward the anode. Both the amplitude and arrival time of the pulse on the sense wires are measured electronically. All the data are recorded on tape and subsequently the coordinates of the ionizing parent particle in space can be reconstructed. The measurement of the drift time can be used to determine the distance of the parent particle from the wire; the distance along the wire is determined by comparing the sizes of pulse received

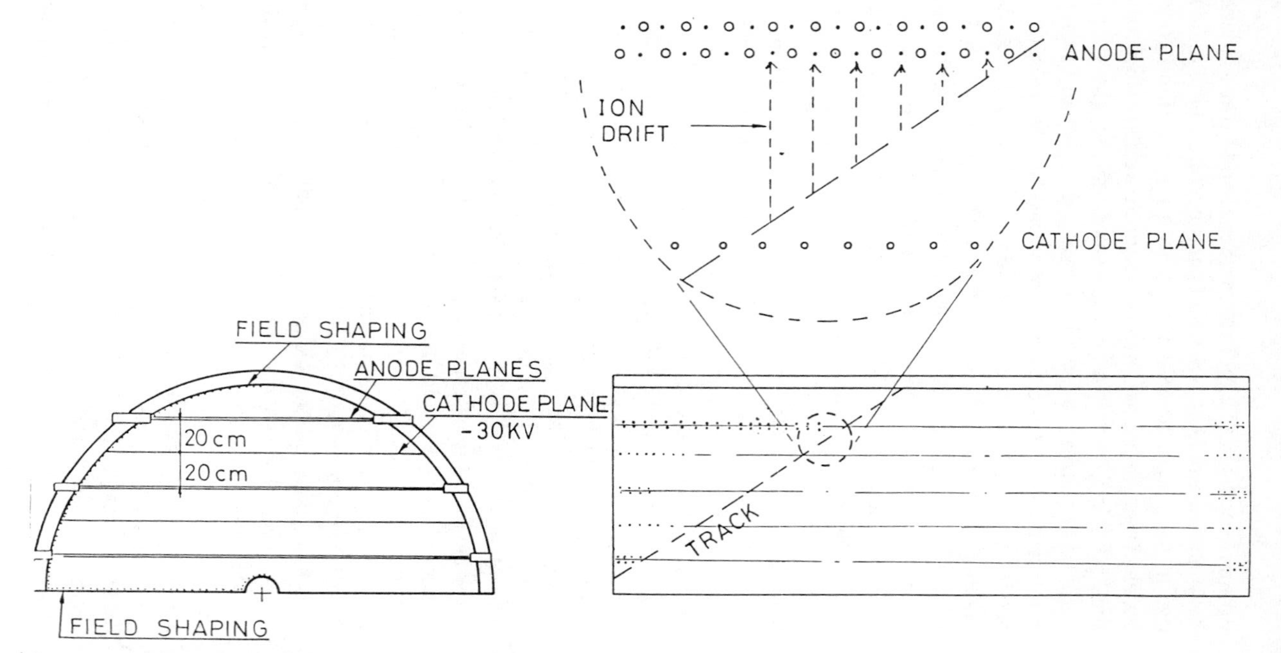

Fig. 3.8 The side and end sections of a drift chamber used in the UA1 experiment.

at the two ends of the wire. Between the anode wires are regions of low field where ions drift slowly and because of this it would be difficult to correlate drift time with the distance of the primary ionization from the wire plane. Therefore thicker field wires held at -2 KV are alternated with the sense wires. The maximum separation of the planes is dictated by the need for all electrons to have drifted to the anode before the next proton and antiproton bunch collide at the crossing point inside the detector. Crossing occurs at intervals of 7·2 μs and so interactions of interest may occur at this rate. Fig. 3.9 shows the side view of an event in which a W^--boson was produced. The electron from the decay

$$W^- \to e^- + \bar{\nu}_e$$

is arrowed. An applied magnetic field (0·7 T) permits the measurement of charged particle momenta. A precision of around 4 ns on timing gives a position accuracy of ~ 300 μm and the momenta are measured to a precision

$$\Delta p/p \simeq 0{\cdot}005\ p,$$

where the momenta are in GeV/c.

3.5 Scintillation counters

These detectors are constructed from materials which scintillate when charged particles travel through them. The light emitted is usually detected by a photomultiplier. Both inorganic and organic scintillators are used. Of the first category a commonly employed material is sodium iodide in the form of a large crystal doped with thallium. The doping is essential because it produces the colour centres which scintillate. Free electrons and holes produced by the passage of ionizing radiation can be captured by a colour centre which will then be in an excited state. The excited state decays with a lifetime of 0·25 μs emitting a visible photon. Organic materials rely on a different mechanism to give light: molecules are excited by ionizing radiation and decay after a very short lifetime (~ 1 ns) with the emission of ultraviolet photons. The active material is dissolved

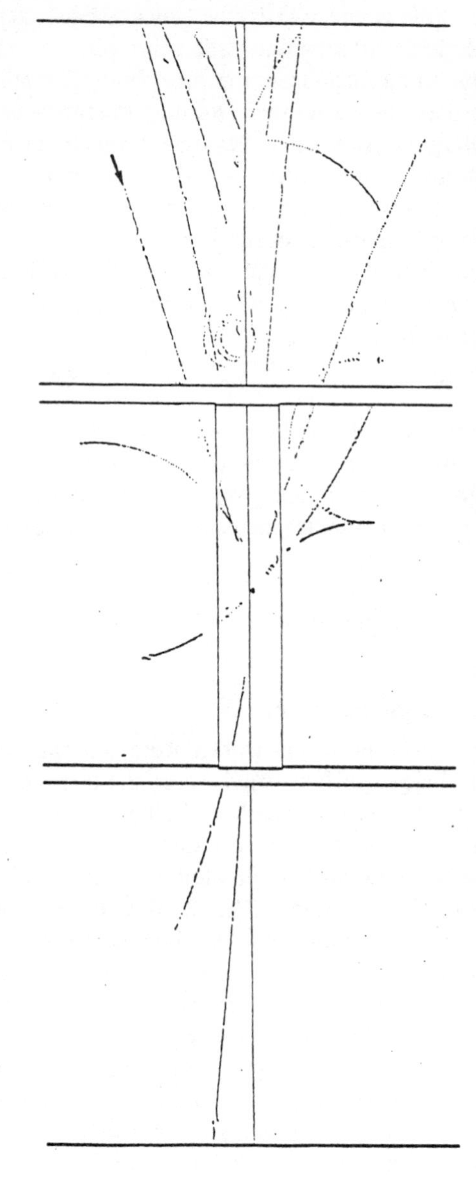

Fig. 3.9. A photograph of the reconstructed tracks observed from a $\bar{p}p$ interaction in the UA1 detector in which a W^--boson is produced and decays. The electron track is arrowed (from G. Arnison *et al.*, *Phys. Lett*. **122B**, 103 (1983)).

in a slab of transparent plastic such as Plexiglass. In order to match the wavelength at which the response of photomultipliers is peaked the wavelength of the radiation needs to be shifted out of the ultraviolet. This is done by doping the Plexiglass with a dye that absorbs in the UV and emits in the visible spectrum. Organic scintillator is very widely used because plastic can easily be cast into large and complex shapes; it also gives better time resolution. Sodium iodide has the advantage of linearity between light output and energy deposition down to MeV energies; however, it is more expensive, has slower response and being hygroscopic it needs careful encapsulation.

The scintillator is viewed by a photomultiplier either directly or via a light guide made of polished plastic. Light is transmitted along these light guides by total internal reflection. There is a large loss of light but it becomes possible to locate the photomultiplier in an accessible and field free place. Photomultipliers have photosensitive cathodes containing alkali metals; these have around 10 per cent quantum efficiency for responding to photons by emitting electrons. Near the cathode are arranged a series of about ten electrodes (dynodes) at progressively more positive voltages with respect to the cathode (e.g. +100 V, +200 V, . . .). These are coated with a good secondary emitter BeO or Mg-O-Cs, which yield 3–5 electrons when struck by an electron of over 100 eV. An electron from the photocathode is accelerated to the first dynode where it liberates electrons, which are accelerated to the second dynode, and so on. An amplification over the chain of dynodes of 10^6 is readily obtained with a 10 ns pulse length which gives millivolt pulses in an external 50 ohms resistance following the last dynode. Scintillation counters can be used as timing and triggering devices because of their good time resolution and detection efficiency. The passage of a beam particle just ahead of the target may be used to define the time of an interaction. A time window can then be applied to data detected by other apparatus; this provides a powerful way to reduce backgrounds.

3.6 Calorimeters

Calorimeters are devices used to measure the total energy of individual particles or collimated bunches (jets) of particles. We must distinguish between electromagnetic calorimeters and hadron calorimeters, the former being designed to measure the energy of photons, electrons and positrons while the latter are intended to measure the energy of hadrons. One form of electromagnetic calorimeter is a sandwich of plates of a material of high atomic number interleaved with sheets of scintillator. The integrated thickness amounts to twenty or more radiation lengths in order that all but a negligible fraction of the photons or electrons incident do initiate electromagnetic showers which are completely absorbed in the calorimeter. When the shower crosses a sheet of scintillator each of its component electrons and positrons excites the emission of light. The signal from the photomultiplier which views this scintillator plate is proportional to the number of particles crossing the sheet. In turn the total output of the photomultipliers viewing all the sheets of scintillator is proportional to the total energy of the shower and hence, as explained in section 3.2, to the energy of the incident particle. The precision of the energy measurement is determined by the statistical fluctuations in the numbers of charged particles crossing the scintillator sheets. For a lead-scintillator calorimeter with ~ 2 mm lead sheets

$$\sigma(E) \simeq 0{\cdot}15\sqrt{E}\ \text{GeV}$$

with E in GeV.

An alternative way to measure the energy in the shower is to collect the primary ionization produced by the positrons and electrons. The sampling medium sandwiched between the lead plates is liquid argon with electrodes immersed in the liquid in each gap. The ionization produced in the liquid is collected on these electrodes and integrated.

Yet another form of electromagnetic calorimeter uses blocks of dense glass loaded with lead in which the Cerenkov light emitted by the electrons is viewed by photomultipliers. This is expensive but achieves a resolution of $\sim 5\%\sqrt{E}$. At lower energies (0·1–0·5 GeV) single thick crystals of sodium iodide can be used to absorb and measure the energy of an electromagnetic shower. The 'crystal ball' detector used at SLAC is a full solid angle (4π) detector of this variety.

Hadron calorimeters can also consist of sandwiches of a heavy material and a detecting medium. In this case the parameter determining the required thickness of material is the nuclear interaction length, which is comparable in iron and lead (17 cm). Five interaction lengths represent a practical limit (85 cm of iron). Energy resolution in a hadron calorimeter is much poorer because the shower fluctuations are more severe. In practice

$$\sigma(E) \approx 0{\cdot}8\sqrt{E}\ \text{GeV}$$

for an iron-scintillator calorimeter with E in GeV.

3.7 The UA1 detector

One large-scale device that incorporates lepton and hadron detection is the UA1 detector shown in Fig. 8.5. It will serve to show the principle of a layered detector, with each layer performing a different task. The UA1 detector was designed to identify and measure the energies of leptons of energies around 40 GeV coming from the vector-boson decays:

$$W^- \to e^- + \bar{\nu}_e \quad \text{or} \quad \mu^- + \bar{\nu}_\mu;$$
$$Z^0 \to e^+ + e^- \quad \text{or} \quad \mu^+ + \mu^-;$$

and

$$W^+ \to e^+ + \nu_e \quad \text{or} \quad \mu^+ + \nu_\mu.$$

Fig. 8.5 (p. 165) shows a vertical section through the apparatus with the evacuated beam pipe running through the centre. Protons inside the pipe coming fom the left meet antiprotons from the right at a point midway through the central detector (see above). The products of a $\bar{p}p$ interaction emerge from the 20 cm diameter beam pipe, cross the central detector and pass into the calorimeters. The aluminium coil provides a field of 0·7 T perpendicular to the diagram, so that measurements of track curvature can be used to determine the momenta of charged particles. Outside the central detector lie in turn the lead-scintillator electromagnetic calorimeter and the iron-scintillator hadron calorimeter. These form closed shells as far as practicable around the interaction point. They are built in cells so that the energy deposition can be localized and each of these cells is sampled in depth (four electromagnetic samplings and two hadronic samplings). In all, several thousand photo-multipliers are used. The electromagnetic calorimeter is 26 radiation lengths thick so that only a fraction, $\exp(-26)$, of the electrons incident cross it without showering. Together the calorimeters are six nuclear interaction lengths thick so that hadrons rarely penetrate beyond. Outside the calorimeter is a layer of planar drift chambers, which identify muons. Muons travel through all the material losing energy by ionization to a total of about 2 GeV. The whole device weighs 1400 tonnes.

An electron will produce a track in the central detector and deposit all its energy in the electromagnetic calorimeter. Electron detection therefore requires the observation of an isolated track in the central detector and a matching energy deposition in the electromagnetic calorimeter but *no* energy deposition in the cells of the hadron calorimeter behind the electromagnetic cells affected. Muons leave a track in the central detector, a small energy deposition in the calorimeter cells traversed and penetrate to the external drift chambers. The presence of these features is therefore demanded for positive muon identification. It is also necessary that the extrapolation of the track from the central detector agrees in angle and position with the hits in the external muon chambers. This last requirement eliminates those hadrons leaking through the material at the end of a hadron shower.

A final word concerns neutrino detection. Neutrinos of course escape without depositing energy, so that detection must be indirect. The calorimeters cover almost the full solid angle around the interaction, so it is possible to make a vector balance of the energy (and hence momentum) detected by the calori-meter cells. This balance is only well determined in directions perpendicular to the beam axis because energetic particles can escape detection by travelling close to the beam axis inside the beam pipe after the interaction. The transverse energy balance is measured to a precision of a few GeV. A neutrino emitted in W-decay which travels transversely is revealed as a missing transverse energy of around 40 GeV, and can be identified cleanly.

3.8 Particle identification

In this section attention is directed to methods for discriminating between charged *hadrons*. Two methods have already been mentioned: Cerenkov radiation detection and ionization measurement. Two other techniques also need comment. They are time-of-flight measurement and transition radiation detection.

Time-of-flight (TOF) is the interval taken by a particle to travel between two detectors. When a particle's momentum is determined from its motion in a magnetic field (e.g. in a drift chamber) the TOF measurement can be used to identify the particle type. Velocities are typically close to c with only weak variation with particle mass; scintillation counters provide the necessary good time resolution. A resolution of 300 ps is achievable and permits discrimination of π- from K-mesons to a momentum 1·5 GeV/c over a flight path of 5 m.

Ionization measurements were mentioned earlier and examples are shown in Fig. 3.6. Precisions of $\sim$10 per cent seem to be attainable from metre-long tracks in drift chambers. This method is useful over the range 0·2–1 GeV/c for π/K separation. Discrimination in the region of the relativistic rise has also been reported.

Cerenkov counters are the most versatile devices for discriminating between particle types, not only in beam lines but as part of a detector. The number of Cerenkov photons emitted in the wavelength range to which a typical photomultiplier is sensitive (400–700 nm), is given by Kleinknecht (1986, p. 127) as

$$N_\gamma = 4{\cdot}9 \times 10^4 \{1 - 1/(\beta^2 n^2)\}\ \mathrm{m}^{-1}.$$

Threshold Cerenkov counters are built with a variety of active media in order to provide particle discrimination over a wide momentum interval. These materials range from gases with refractive indices close to unity to glasses with refractive indices of up to 1·75. Threshold Cerenkov counters are used to discriminate π-mesons from K-mesons up to momenta of ~ 20 GeV/c. At higher momenta differential Cerenkov counters are available. In these devices an aperture is used to select Cerenkov radiation emitted at a particular angle. With this refinement it is possible to distinguish π-mesons from K-mesons up to ~ 400 GeV/c.

A more recently developed method of particle identification relies on what is called transition radiation. Transition radiation is emitted when a charged particle crosses the boundary between two media with differing dielectric constants. It is concentrated in a cone of opening angle γ^{-1}. A stack of thin foils and spacers can give a coherent effect: an early example used one thousand 50 μm lithium foils with a xenon proportional counter employed to detect the transition radiation X-rays. This technique has promise for resolution at γ-values exceeding 1000, a regime in which there are no competitive methods.

3.9 Detectors for neutrino interactions

These devices are necessarily massive because the neutrino cross-section is so small. At 100 GeV the total cross-section is 10^{-40} m^2/nucleon, so that with 10^9 neutrinos incident in a narrow-band beam (Fig. 3.5) the number of interactions per pulse in 1 m of iron is only

$$N = 10^{-40} \times 10^9 N_0 \rho,$$

where N_0 is Avogadro's number ((kg mole)$^{-1}$) and ρ is the density (kg m^{-3}), i.e.

$$N = 10^{-31} \times 6{\cdot}02 \times 10^{26} \times 7{\cdot}87 \times 10^3 \approx 0{\cdot}5.$$

Neutrino interactions have been studied using bubble chambers filled with heavy liquid. These devices have the advantage that the individual tracks emerging from an interaction can be observed and this advantage proved invaluable in recognizing neutral-current processes (see section 8.3). However, the target mass and hence event rate is low. Experiments with calorimeters achieve a larger event rate but yield less detail. The Cern-Dortmund-Heidelberg-Saclay (CDHS) collaboration used a calorimeter at CERN to accumulate definitive data on nucleon structure discussed in section 9.2. Their calorimeter consists of a stack of 19 toroidal iron magnets each with inner/outer diameter 0·2/3·75 m and of thickness 0·75 m in the beam direction. Coils wound on the toroids provide a 1·65 tesla field. The front seven magnets form a target plus calorimeter having scintillator sheets every 5 cm of iron to sample the showers. The remaining magnets have some scintillator sampling, but are primarily used in determining the track curvature and hence momentum of any μ-leptons emerging from the neutrino interactions. Large-area drift chambers are located between each of the magnet modules to record particle tracks. The complete device weighs 1500 tonnes.

Chapter 4
QED and the gauge principle

The quantum field theory of electromagnetic processes is called quantum electrodynamics (QED) and is the simplest of the gauge theories. It has been used to make precise predictions which have been checked to the limit of experimental accuracy. In this chapter we start by describing the relativistic quantum theory of the electron due to Dirac. Next the principles of quantum electrodynamics are introduced and an account given of Feynman diagrams. These diagrams are helpful in visualizing electromagnetic processes and in calculating reaction amplitudes. The process of e^+e^- annihilation provides an application of the techniques which is of relevance to the study of quark properties. The above sections are supplemented by more detailed mathematical material in Appendices D and E. In a final section the underlying significance of the gauge principle to QED is considered at length.

4.1 Dirac's theory of the electron

Dirac developed a theory of the electron which is relativistic and from which the existence of its antiparticle—the positron—emerges naturally. The non-relativistic spin-up and spin-down states of the electron are represented by two-component entities called spinors (ϕ_s):

$$\phi_+ = \begin{bmatrix} 1 \\ 0 \end{bmatrix} \quad \phi_- = \begin{bmatrix} 0 \\ 1 \end{bmatrix}.$$

In the Dirac theory there are four basic states of the electron: two with positive energy, one having spin up and the other having spin down; and two with negative energy. Relativistic electrons are therefore described by four component spinors $u(p, s)$ where p is the four-momentum and s is the third component of the spin:

$$u(p, s) = \sqrt{E+m}\begin{bmatrix} \phi_s \\ (\boldsymbol{\sigma} \cdot \mathbf{p})\phi_s/(E+m) \end{bmatrix}$$

where $\phi_s = \phi_\pm$ in the case that $s = \pm\frac{1}{2}$ respectively. $\mathbf{p}$ is the three-momentum and $\boldsymbol{\sigma}$ is the Pauli spin matrix. More details are given in Appendix D. Then the wavefunction of a free electron with four-momentum p_μ is the product of a plane wave and a spinor:

$$\psi(p, s) = u(p, s)\exp(-ip_\mu x_\mu).$$

The Dirac wavefunctions are the solutions of the Dirac equation for a free electron:

$$(i\gamma_\mu \partial_\mu - m)\psi = 0$$

where we recognize $i\partial_\mu$ as the four-momentum operator discussed in section 1.1. The component γ's form a four-vector in space-time, but have a more complex structure than vectors. Each component is itself a four-by-four matrix acting in the 'spin' space spanned by the four-component spinors u. By implication the mass in the Dirac equation needs to act on the spinors: m is taken to mean mI_4, where I_4 is the four-by-four unit matrix with 1's along the diagonal and zeros elsewhere.

Dirac accounted for the negative-energy states in the following way. He suggested that in the vacuum all negative-energy states were normally filled; because of the Pauli Exclusion Principle this arrangement would prevent positive-energy electrons dropping into the negative-energy states. The vacuum taken as the reference for energy would in fact be biased to an enormous negative energy. If an electron were excited from this sea of negative-energy states to positive energy it would leave a hole in the negative-energy distribution. The empty negative state would be available for another electron to enter. In such cases the motion of the hole is opposite to that of the electron. It would behave like a positively charged particle of the same mass as the electron. Dirac regarded a hole as the antiparticle of the electron, which is called the positron. This interpretation was shown to have some validity when in 1932 Anderson discovered the positron.

Dirac's idea fails when applied to bosons. Bosons do not obey the Pauli exclusion principle so the continuum of negative-energy states can never be filled. Quantum electrodynamics gives a more consistent explanation which removes the need for a continuum of occupied negative-energy states. In the language used by Feynman a negative-energy particle moving backward in time is exactly equivalent to a positive-energy antiparticle moving forward in time. The positron is thus equivalent to a negative-energy electron going backward in time.

In the presence of an electromagnetic field $A = (A_0, \mathbf{A})$ the standard classical expression for the energy of a particle with charge q and mass m is:

$$E = (1/2m)(\mathbf{p} - q\mathbf{A})^2 + qA_0$$

i.e.

$$(E - qA_0) = (1/2m)(\mathbf{p} - q\mathbf{A})^2.$$

The inference can be made that the four-momentum p is replaced by $p - qA \equiv (E - qA_0, \mathbf{p} - q\mathbf{A})$. This expresses what is called the minimal electromagnetic coupling. The Dirac equation for an electron in the presence of an electromagnetic field becomes, as a consequence,

$$[\gamma_\mu p_\mu + e\gamma_\mu A_\mu - m]\psi = 0.$$

p_μ must be rewritten in its operator form, giving

$$(i\gamma_\mu \partial_\mu - m)\psi = -e\gamma_\mu A_\mu \psi.$$

Neutrinos are also spin-$\frac{1}{2}$ fermions but unlike the electrons possess no mass or charge. In Appendix D it is shown that only a two-component spinor w is needed to describe the neutrino and that this satisfies the equation

$$Ew(v) = -\boldsymbol{\sigma} \cdot \mathbf{p}w(v).$$

Now $E = |\mathbf{p}|$ for a massless particle so that $\boldsymbol{\sigma} \cdot \mathbf{p}/E$ is the spin component along the direction of motion and has eigenvalues ± 1; this spin component is called the helicity (λ). The neutrino has negative helicity ($\lambda = -1$) i.e. it is *left-handed*. Its antiparticle the antineutrino is also massless and is *right-handed*. The sign of the antineutrino helicity can be deduced with the help of Dirac's interpretation of negative-energy states. When a negative-energy neutrino moves to fill a hole the hole moves in the opposite sense. Compared with the neutrino the hole has its momentum reversed and spin direction in space reversed; the hole also has positive energy. Thus $\boldsymbol{\sigma} \cdot \mathbf{p}/E$ is reversed; antineutrinos are therefore right-handed if neutrinos are left-handed. We can observe that the operators

$$P_{\mathrm{L}} = \tfrac{1}{2}(1 - \boldsymbol{\sigma} \cdot \mathbf{p}/E) \quad \text{and} \quad P_{\mathrm{R}} = \tfrac{1}{2}(1 + \boldsymbol{\sigma} \cdot \mathbf{p}/E)$$

project out the left- and right-handed components of two-component spinors. The appropriate forms to act on four-component spinors are given in Appendix D, namely

$$P_{\mathrm{L}} = \tfrac{1}{2}(1 - \gamma_5) \quad \text{and} \quad P_{\mathrm{R}} = \tfrac{1}{2}(1 + \gamma_5).$$

This analysis shows that because they are massless the neutrinos which exist in nature have a definite handedness. We shall see in Chapter 5 that this has been experimentally verified. Note that all massless particles seem to lose spin components; the photon may have helicity $+1$ (right-circularly polarized) or -1 (left-circularly polarized) but never zero.

4.2 Feynman diagrams

The modern view of the interactions of elementary charged particles and the electromagnetic field is embodied in a quantum field theory: quantum electrodynamics.

Quantum electrodynamics was independently developed by Feynman (1948), Tomonaga (1948) and Schwinger (1948). Feynman introduced diagrams that go by his name to illustrate the interactions and we shall stick to this way of viewing electromagnetic processes. The diagrams are built from the interaction vertex shown in Fig. 4.1, where the straight lines

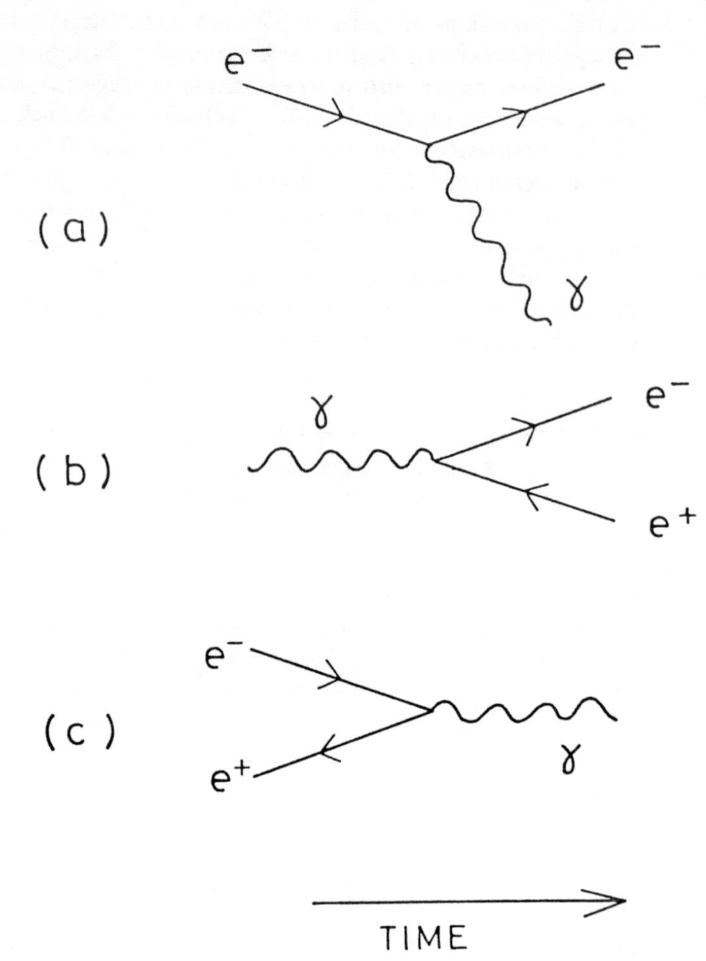

Fig. 4.1 Vertices for electromagnetic coupling. (a) The emission of a photon by an electron. (b) The conversion of a photon to an electron-positron pair. (c) The annihilation of a positron on an electron.

indicate the electrons or positrons and the wavy line indicates a photon. Time flows to the right and the arrows indicate the electron direction in time. Fig. 4.1(a) shows an electron emitting a photon. Fig. 4.1(b) shows a photon converting to an electron-positron pair, where we see that the arrowhead points backward in time for the positron. Finally Fig. 4.1(c) shows an electron-positron annihilation. These diagrams bring out the underlying connection between the three processes; the similarity goes further because the same amplitude describes all three processes.

Some Feynman diagrams for processes contributing to electron–electron scattering are shown in Fig. 4.2. Fig. 4.2(a) is the simplest and most important diagram: a single 'virtual' photon is emitted by one electron and absorbed by another electron. The quantum mechanics of such a process were outlined in section 2.4. In Fig. 4.2(b) and 4.2(c) an additional virtual photon is emitted and reabsorbed by the same electron. In Fig. 4.2(d) the exchanged photon converts to a pair which later annihilate. Finally in Fig. 4.2(e) two photons are exchanged.

The reader can easily picture more complicated diagrams leading to ones with infinite numbers of vertices. The totality of such diagrams describes all possible ways to get from the initial state (incoming electrons) to the final state (outgoing electrons). The quantum mechanical amplitude, T, for getting from the initial to final state is the coherent sum of the amplitudes of all these diagrams. What makes this approach useful for calculations is that the simplest diagrams make the dominant contribution to the amplitude. Each vertex involves the electromagnetic coupling once and brings into the (amplitude)^2 a factor $\alpha(1/137)$. Therefore the diagrams form a perturbation series where each additional internal photon or electron-positron loop reduces the numerical importance of the contribution of the diagram to the (amplitude)^2 by a factor α^2. The convergence is rapid so that for most calculations only a few diagrams of low order in α need be considered.

In terms of the amplitude T the differential cross-section for a two-body to n-body reaction is:

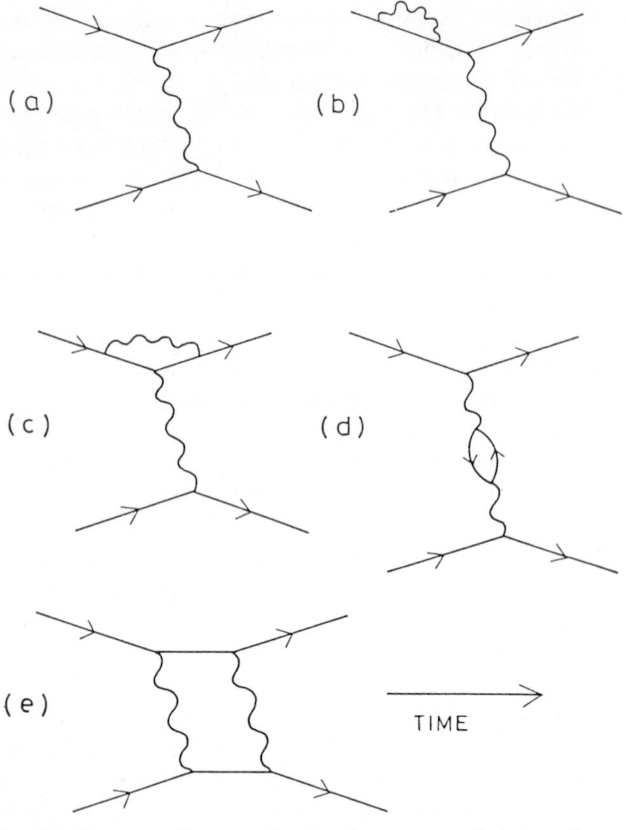

Fig. 4.2 Feynman diagrams for the electromagnetic interactions of two electrons.

$$d\sigma = |T|^2 (dQ_n/K)/F,$$

where dQ_n/K is the number of quantum mechanically distinguishable states in an n-body final state, K being a statistical factor equal to $k!$ when there are k identical particles in the final state. F is the flux of initial-state particles used in calculating T. T, dQ_n and F are all Lorentz-invariant quantities. Evaluation of these factors for a two-body final state is described in Appendix A and gives in the CM frame

$$dQ_n/F = d\Omega^*/(64\pi^2 s),$$

where s is the centre-of-mass energy squared and $d\Omega^*$ is the element of solid angle defined by the detector.

Feynman's rules for evaluating individual amplitudes are based on the diagrams and here a few explanatory remarks are made on the scheme. For external lines the rules are to write the spinor for each fermion and the polarization vector for each photon. Each vertex contributes a factor $-ie\gamma_\mu$ which is derived directly from the form of the minimal electromagnetic interaction. This can be seen, for example, in the Dirac equation describing an electron in the presence of an electromagnetic field. The internal lines contribute factors $i/(q_\mu\gamma_\mu - m)$ for fermions, i/q^2 for photons (discussed in section 1.4) and $1/(q^2 - m^2)$ for massive bosons. These so-called propagator terms which arise from an exchange are most easily understood by considering the emission of a particle of four-momentum q_μ from a fixed source Q. For the case of electron exchange the Dirac equation becomes:

$$(i\gamma_\mu\partial_\mu - m)\psi = Q.$$

The electron wavefunction is a plane wave

$$\psi = u \exp(-iq_\mu x_\mu)$$

and substitution gives

$$(\gamma_\mu q_\mu - m)\psi = Q.$$

Therefore:

$$\psi = Q/(\gamma_\mu q_\mu - m).$$

In the case that a boson of mass m is exchanged we use the Klein-Gordon equation (Appendix D):

$$(-\partial_\mu\partial_\mu - m^2)\phi = Q.$$

Substitution of a plane wave for ϕ gives

$$\phi = Q/(q^2 - m^2).$$

For a massless boson, the photon, we get a propagator of the form $1/q^2$. This latter was precisely the form of the amplitude for Coulomb scattering found in section 2.4, thus confirming the reasonableness of the approach. The propagator for massive-boson exchange will be needed later when discussing the weak interaction. The four-momenta of particles forming the internal lines of diagrams are constrained only by the requirements of conservation of four-momentum at each vertex. All such particles are clearly virtual.

Calculation of an amplitude will in general involve integrations over all the four-momenta of the internal lines. Difficulties occur in evaluating the contribution of the loop shown in Fig. 4.2(d). Suppose the photon four-momentum is k and the electron and positron momenta in the loop are p and $(k-p)$. Then the piece of the amplitude coming from the loop is

$$\int \mathrm{d}^4p/[\gamma(p-k)-m][\gamma p-m] \sim \int \mathrm{d}^4p/p^2$$

which diverges as p tends to infinity because the volume element d^4p grows faster than p^2. The solution to this and similar difficulties lies in appreciating that the lines in Feynman diagrams represent bare electrons and not the real (physical) electrons observed in the laboratory. Real electrons are continuously emitting and reabsorbing virtual photons and *cannot* be detached from this accompanying cloud of virtual photons. Technically all the divergences in all diagrams whatever can be absorbed into just two constants, which are the difference in mass (Δm) between the real (m) and bare (m_0) electron, and the corresponding charge difference (Δe):

$$m = m_0 + \Delta m$$
$$e = e_0 + \Delta e.$$

Then m_0, e_0, Δm and Δe are all infinite but unobservable while e and m remain finite. This procedure is called renormalization of the charge and mass. A theory that gives finite reaction rates and observables at the cost of a restricted number of unobservable infinite constants is called renormalizable. The property of renormalizability coupled with the small size of fine structure constant means that it is possible to make predictions of high accuracy with QED whilst only calculating the contribution of relatively few low-order diagrams.

Corrections to the simple Dirac theory are significant. According to Dirac's theory the magnetic moment of the electron is given in terms of the spin **s** by

$$\boldsymbol{\mu}=g(e/2m)\mathbf{s}$$

with a Landé g-factor value of exactly 2. This value would be correct if the interaction of an electron with an external magnetic field involved a single-photon exchange as in Fig. 4.1(a). However, a diagram in which the electron radiates a photon before the interaction with the external field, and then reabsorbs this photon later also contributes. This and more involved diagrams give 'radiative corrections' which lift g to a value of 2·00232. Measurements of $(g-2)$ for both electrons and muons confirm the QED predictions to a few parts in 10^{10}; an achievement which has demanded the calculation of contributions of some hundred diagrams on the one hand and refined experiments (see Chapter 2) on the other.

The diagram shown in Fig. 4.2(d) is one of those responsible for what is termed 'vacuum polarization'. In general an electron emits photons which can convert to electron-positron pairs. The parent will repel the electron and attract the positron of each pair, inducing a polarization of the vacuum which serves to shield the bare electron charge (e_0). Millikan's oil-drop experiment measures the static charge (e). If instead fast electrons are used to probe an electron they can approach more closely to the electron and the shielding due to vacuum polarization becomes less effective. The observed charge therefore tends towards the value of the bare charge, i.e. it increases. We can make use of the uncertainty principle to link the momentum transfer (q) and the closest distance of approach (b): $qb\sim h$. Thus the value of measured charge is expected to increase as b decreases or equivalently as q increases. In detail, in terms of the Lorentz-invariant quantity q^2:

$$\alpha(q^2)=\alpha(m^2)\left[1-(\alpha(m^2)/3\pi)\ln(|q^2|/m^2)\right]^{-1}$$

for $-q^2\gg m^2$, where $\alpha(m^2)$ is the static value of the fine-structure constant α measured in Millikan's experiment. We have here a very surprising result: the electric charge of the electron is *not* a constant but it depends on the four-momentum transfer made to the electron in the measurement process.

The variation of α with q^2 leads to first-order shifts in atomic energy levels. In hydrogen the electron Bohr radius is $(m\alpha)^{-1}$ and the typical momentum exchange q is therefore only αm, so that the effect is small. The $2s_{\frac{1}{2}}$ and $2p_{\frac{1}{2}}$ levels would be degenerate in the Dirac theory: however, the vacuum polarization raises the $2s_{\frac{1}{2}}$ level by 27 MHz (0·11 μeV). This is a measurable effect although swamped by other radiative effects, which raise the $2p_{\frac{1}{2}}$ level by a total of +1057·9 MHz above the $2s_{\frac{1}{2}}$ level. This total displacement is called the Lamb shift and was measured and calculated by Lamb and Rutherford (1947). In a muonic hydrogen atom where a massive muon replaces the electron the Bohr radius is smaller by a factor m_μ/m so that (q/m) is of order unity. The main contribution to the Lamb shift now comes from vacuum polarization with the $2s_{\frac{1}{2}}$ level lying higher than the $2p_{\frac{1}{2}}$ level by 202·0 meV. Levels in heavier muonic atoms of atomic number Z show more striking effects because the Bohr radius is even smaller $(Zm_\mu\alpha)^{-1}$ and Lamb shifts of energy levels of up to 100 eV are observed (Borie and Rinker (1982)).

Single-photon exchange dominates the electromagnetic processes by virtue of the fact that higher-order diagrams are suppressed by powers of α (1/137). The calculation of the amplitude and cross-section for the elastic scattering of point-like spin-$\frac{1}{2}$ fermions $1+2\rightarrow 3+4$ is given in detail in Appendix E. Using the Feynman rules quoted above the amplitude for single-photon exchange is:

$$T=[\bar{u}(k_3, s_3)\gamma_\mu u(k_1, s_1)]\,\frac{e^2}{q^2}\,[\bar{u}(k_4, s_4)\gamma_\mu u(k_2, s_2)].$$

Each term in square brackets is called a current: it contains a spinor for the incoming particle and an adjoint spinor for the outgoing particle (see Appendix D). Between the currents is sandwiched the propagator for the exchanged photon with four-momentum q equal to $(k_1 - k_3)$. What precisely is the significance of γ_μ? A current containing γ_μ is a vector in space-time and matches the vector nature of the photon. Each component γ_μ is a 4×4 matrix acting on the spinors. Put crudely $\gamma_\mu u(k_1 s_1)$ describes the spin state of a fermion which absorbs a photon. Taking the scalar product $\bar{u}(k_3, s_3)\gamma_\mu u(k_1, s_1)$ gives the spinor overlap with the outgoing state of interest, namely, $\bar{u}(k_3, s_3)$. By having the same γ_μ in both currents we ensure that the polarization of the photon emerging from one vertex matches its polarization when it is absorbed at the other vertex, as indeed it should.

4.3 Electron-positron annihilation

The annihilation of positrons on electrons in a colliding-beam experiment produces a single photon. If the electron and positron have four-momenta $(E, \mathbf{p})$ and $(E, -\mathbf{p})$ then the photon has a four-momentum $(2E, 0)$; it has energy but no momentum! These are again virtual photons, and differ from real photons (in say a beam of γ's) which have $E = p$. Departures from this equality are only possible thanks to the uncertainty principle: an energy imbalance $2E$ can exist for a length of time $t \sim h/2E$. So the virtual photon must convert in time t. A number of options are available, one of which gives the sequence

$$e^+ + e^- \rightarrow \gamma \rightarrow \mu^+ + \mu^-. \qquad [a]$$

This process is compared to the reaction

$$e^- + \mu^- \rightarrow e^- + \mu^- \qquad [b]$$

in Fig. 4.3: we see that [b] is obtained from [a] by the replacement

μ^+ (outgoing) $\rightarrow \mu^-$ (incoming)

e^+ (incoming) $\rightarrow e^-$ (outgoing).

The processes are therefore equivalent when it comes to writing the Feynman amplitude. This feature is made use of in Appendix E where the differential cross-section for $e^+ + e^- \rightarrow \mu^+ + \mu^-$ is calculated. The result in the CM frame is

$$d\sigma/d\Omega^* = (\alpha^2/16E^2)(1 + \cos^2\theta^*)$$

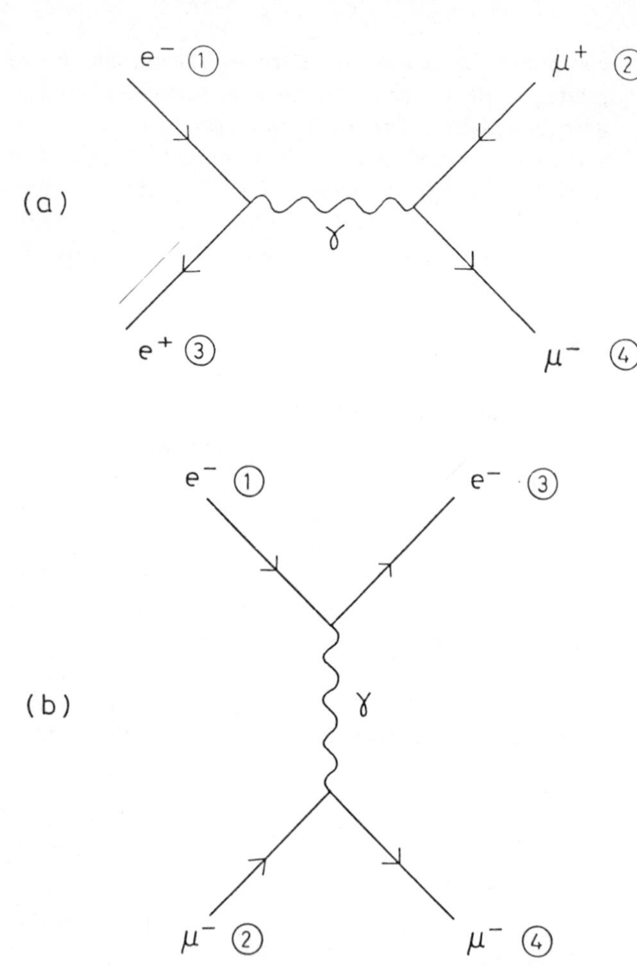

Fig. 4.3 Feynman diagrams for: (a) $e^- + e^+ \to \mu^- + \mu^+$; (b) $e^- + \mu^- \to e^- + \mu^-$.

where $d\Omega^* = 2\pi d(\cos\theta^*)$ is the element of solid angle into which the μ^--lepton emerges, θ^* is the angle it makes with the direction of the incident electron and E is the beam energy. This result is valid provided that the beam energy is well above the threshold for producing the lepton pair. In such circumstances the measured differential cross-section for μ-lepton pairs (or

τ-lepton pairs) agrees excellently with the prediction. When the differential cross-section is integrated over solid angle this gives the total cross-section for the process:

$$\sigma = 4\pi\alpha^2/3s,$$

where $\sqrt{s} = 2E$ is the CM energy.

A significant fraction of annihilations gives final states containing hadrons only and no leptons. According to the quark model these final states are mediated by the process

$$e^+ + e^- \rightarrow \gamma \rightarrow q + \bar{q}$$

where subsequently the quark and antiquark fragment into hadrons with unit probability because we never observe free quarks. At high enough energies the hadrons from the parent quark (or antiquark) show up as a collimated bunch of hadrons, known as a jet. Fig. 4.4 shows an event observed in a drift chamber surrounding the e^+e^- interaction point, where the two back-to-back jets are quite obvious. The view is a projection along the beam axis; the curvature of the particle trajectories in the axial magnetic field leads to their separation as they move away from the interaction point (vertex). Because the underlying process for hadron production is again an electromagnetic process involving point-like spin-$\frac{1}{2}$ fermions the differential cross-section must be the same as that for lepton pair production:

$$d\sigma/d\Omega^* = (\alpha^2/16E^2)(1 + \cos^2\theta^*)\,\Sigma Q^2$$

where ΣQ^2 is the sum over the squares of charges for the different quark species. If the quark spin were different from $\frac{1}{2}$ the predicted angular distribution would change. For example, if the quark spin were zero then $(1 + \cos^2\theta^*)$ would be replaced by $(1 - \cos^2\theta^*)$. The data agree excellently with the prediction for spin-$\frac{1}{2}$ quarks (Hanson *et al.* (1975)).

We note that the annihilation (or direct channel) photon of Fig. 4.3(a) has $E^2 - \mathbf{p}^2 > 0$, while the exchanged photon of Fig. 4.3(b) has $E^2 - \mathbf{p}^2 = q^2 < 0$. By analogy with the labelling of space-time separations a direct-channel photon is called 'time-

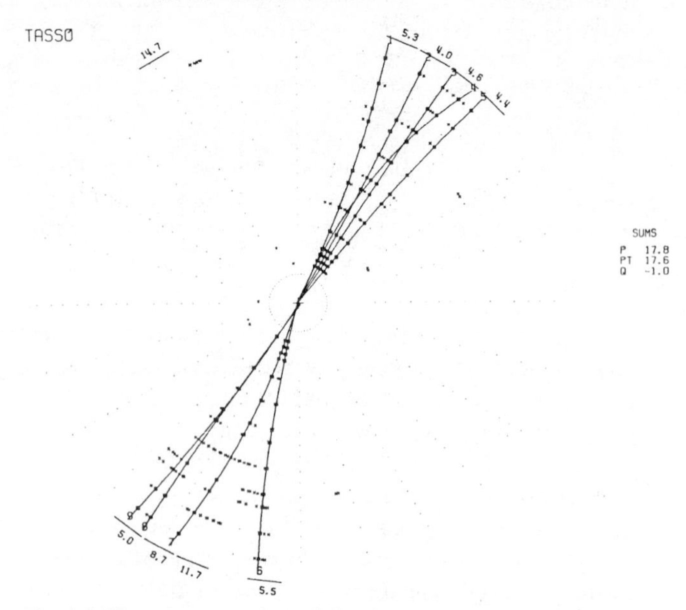

Fig. 4.4 The reconstruction of the charged hadron trajectories from an e^+e^- annihilation showing two back-to-back jets of hadrons. (Photograph courtesy Dr R. Devenish.)

like' and an exchanged photon is called 'space-like'. A real photon is evidently 'light-like'.

4.4 Local gauge invariance

Local gauge invariance is the common feature underlying all the forces of nature: the electromagnetic, weak, strong and gravitational forces. It has also been possible to unify the description of the electromagnetic and weak force using the principles of local gauge invariance (see Chapter 8). Gauge invariance means that it is possible to express a theory in a form which is invariant under a group of transformations made on internal coordinates (for the electroweak and strong forces).

Global gauge invariance refers to independence of the theory to changes of the internal coordinates which are the same at all points in space-time. Local gauge invariance requires independence of the theory under transformations of the internal coordinates which vary smoothly from point to point in space-time. Each force is connected to a particular symmetry of nature and hence to a particular group of transformations and to particular conservation laws.

The simplest gauge symmetry is that connected with charge conservation and this underlies the electromagnetic force and its quantum theory QED. Globally, charge is conserved, $\Sigma q = \text{constant}$: then if each particle wavefunction is multiplied by a factor $\exp(i\alpha q)$ where α is a universal constant, it simply adds a constant phase $\alpha\Sigma q$ to the wavefunction of the universe. For a single particle the effect is

$$\psi \rightarrow \psi \exp(i\alpha q)$$

which is a rotation in some 'internal' space unconnected with space-time. This space has one complex dimension, like an Argand diagram, and because the rotation is determined by the particle charge we may call this space 'charge space'. We can see that a global transformation does not affect any observable. Particle motion is determined by the expectation values of the momentum and energy variables $\psi^*\nabla\psi$ and $\psi^*\dot{\psi}$ which are unaffected by the global gauge transformation

$$\psi^*\partial_\mu\psi \rightarrow \psi^*\partial_\mu\psi.$$

Charge is not only conserved globally, it is also conserved locally in the sense that charge cannot instantaneously move a finite distance in any relativistic theory. This suggests that it would be fruitful to consider a local gauge transformation:

$$\psi \rightarrow \psi \exp(i\alpha(\mathbf{x}, t)q),$$

where α changes smoothly from point to point in space-time. A local gauge transformation leads to a corresponding change in four-momentum:

$$\psi^*\partial_\mu\psi \rightarrow \psi^*\partial_\mu\psi + iq|\psi|^2\partial_\mu\alpha.$$

The momentum can only be made invariant under local gauge transformations by compensating the derivative so that it becomes what is called a covariant derivative. The covariant derivative is

$$D_\mu = \partial_\mu + iqA_\mu(x),$$

where $A_\mu(x)$ is a function of space time (x) which must transform in a special way under local gauge transformations:

$$A_\mu \to A_\mu - \partial_\mu \alpha.$$

Then

$$\begin{aligned} \psi^* D_\mu \psi &= \psi^* \partial_\mu \psi + iq\psi^* A_\mu \psi \\ &\to \psi^* \partial_\mu \psi + iq|\psi|^2 \partial_\mu \alpha + iq\psi^* A_\mu \psi - iq|\psi|^2 \partial_\mu \alpha \\ &= \psi^* D_\mu \psi, \end{aligned}$$

which shows that the momentum variable built from the covariant derivative is invariant under local gauge transformations. What does this procedure mean physically? Firstly, A_μ can be described as a field because it has values at all points of space-time.

More importantly, A_μ can be shown to be consistent with Maxwell's equations, and so may be identified with the electromagnetic field. First construct the tensor

$$F_{\mu\nu} = \partial_\mu A_\nu - \partial_\nu A_\mu$$

which is easily shown to be gauge invariant. Differentiating twice gives

$$\partial_\nu \partial_\mu F_{\mu\nu} = \partial_\nu \partial_\mu \partial_\mu A_\nu - \partial_\nu \partial_\mu \partial_\nu A_\mu = 0.$$

This relation can be compared with the conservation law for electric charge

$$\partial\rho/\partial t + \nabla \cdot \mathbf{j} = 0.$$

where the charge (ρ) and current (j) form a four-vector j_ν. In terms of j_ν this continuity equation becomes

$$\partial_\nu j_\nu = 0.$$

Note that j_ν and $\partial_\mu F_{\mu\nu}$ are both divergenceless. Thus the simplest way to relate the gauge compensating field A_μ to the charge from which it derives its existence is to set

$$\partial_\mu F_{\mu\nu} = j_\nu. \qquad [4.1]$$

This is the equation that was aimed for; it is a compact way of writing Maxwell's inhomogenous equations. A_μ is therefore the four-vector electromagnetic field, in terms of which the electric and magnetic fields are

$$\mathbf{E} = -\partial \mathbf{A}/\partial t - \nabla A_0 \quad \text{and} \quad \mathbf{B} = \nabla \times \mathbf{A}.$$

In covariant notation

$$E_i = \partial_0 A_i + \partial_i A_0 \quad \text{and} \quad B_i = -\partial_j A_k + \partial_k A_j,$$

where i, j and k are taken in cyclic order from 1, 2, 3.

Recall that $\partial_0 = \partial/\partial t$ but $\partial_i = -\partial/\partial x_i$. Then from the definition of $F_{\mu\nu}$,

$$F_{\mu\nu} = \begin{bmatrix} 0 & -E_1 & -E_2 & -E_3 \\ E_1 & 0 & -B_3 & B_2 \\ E_2 & B_3 & 0 & -B_1 \\ E_3 & -B_2 & B_1 & 0 \end{bmatrix}.$$

We can now verify the assertion that we have two of Maxwell's equations. Take the time component of equation [4.1] first,

$$\partial_\nu F_{\nu 0} = j_0$$

i.e.

$$-\partial_1 E_1 - \partial_2 E_2 - \partial_3 E_3 = j_0$$

i.e.

$$\nabla \cdot \mathbf{E} = j_0.$$

From the 1-component of equation [4.1] we get

$$\partial_0 F_{01} - \partial_2 F_{21} - \partial_3 F_{31} = j_1$$

i.e.

$$-\partial_0 E_1 - \partial_2 B_3 + \partial_3 B_2 = j_1.$$

Therefore

$$-\partial \mathbf{E}/\partial t + \nabla \times \mathbf{B} = \mathbf{j}.$$

$F_{\mu\nu}$ automatically satisfies an identity,

$$\partial_\lambda F_{\mu\nu} + \partial_\mu F_{\nu\lambda} + \partial_\nu F_{\lambda\mu} = 0.$$

This contains Maxwell's two homogenous equations,

$$\nabla \cdot \mathbf{B} = 0 \quad \text{and} \quad \nabla \times \mathbf{E} + \partial \mathbf{B}/\partial t = 0.$$

The fields **E** and **B** are directly measurable unlike A_μ and the reader may wish to verify that a local gauge transformation does *not* alter **E** or **B**. This result illustrates an important general principle; that gauge transformations must not affect any observable quantities.

Next we can examine the form of the energy density of the field A_μ. Classically the energy density contains terms quadratic in A_μ or its derivatives only. In addition the energy density is measureable, so it must be gauge invariant also. The standard expression for the energy of the electromagnetic field

$$-\tfrac{1}{4} F_{\nu\mu} F_{\mu\nu} = (E^2 + B^2)/2$$

satisfies both requirements. A term such as

$$m A_\mu A_\mu$$

is feasible and would make a contribution to the energy of a single photon of

$$\int m A_\mu A_\mu \, \mathrm{d}V = m.$$

This would assign a finite mass, m, to the photon. However, the hypothetical expression in $A_\mu A_\mu$ is *not* gauge invariant and so must vanish. Masslessness of the photon is therefore guaranteed by local gauge invariance. Finally, note that the form of the covariant derivative is such that the momentum P_μ of an electron is replaced by $P_\mu + eA_\mu$ in the presence of an electromagnetic field A_μ. This behaviour mimics precisely the form of the minimal electromagnetic coupling described above in section 4.1.

To summarize: the existence of a conservation law (of charge) has led to the requirement of invariance under a group of local gauge transformations and to the need for a gauge field (electromagnetic field) which has massless quanta (photons) and which is coupled in a unique way (minimal electromagnetic coupling) to charge. Local gauge invariance is thus the key to the existence of electromagnetism and its quantized field theory. The gauge transformations take the form of a group of rotations $\alpha(x, t)$ in the Argand diagram for ψ. This internal space can be called 'charge space' and the group is a unitary group U(1). Unitary refers to the property that the transformations do not change the magnitude of $|\psi|^2$ and the 1 refers to the rotation being in one complex dimension.

The success of this procedure suggests that other forces may have a gauge origin. If the relevant symmetry can be inferred, then, by imposing local gauge invariance, the form of the covariant derivative and the gauge field will be determined. This in turn gives the coupling of the field to material particles. What is also true but is not obvious from the example of electromagnetism is that the quarks and leptons should form a fundamental representation of such a group. In other words, whenever an n-fold symmetry exists then the elementary particles should lie in n-fold multiplets whose members possess similar properties. The next chapter will be used to present material about symmetries and groups in more detail.

Chapter 5
Symmetries and conservation laws

In this chapter the properties of both space-time and internal symmetries are discussed. The continuous transformation groups of interest are all of the type called Lie groups. Of these the rotation group is a familiar example and provides a convenient vehicle for bringing out points of general relevance in group theory. Continuous space-time symmetries consistent with Einstein's special theory of relativity are clearly important: in fact they determine what (space-time) quantum numbers a particle may possess. After these, attention turns to the internal symmetries. It is symmetries of this sort which are fundamental to our understanding of the forces of nature. Finally, there are the discrete symmetries such as parity and charge conjugation: these are important paradoxically because they are violated by the weak force.

5.1 Quantum requirements

According to the superposition principle any single-state wavefunction can be expressed as a linear sum

$$\psi = \sum_{i=1}^{n} a_i \phi_i,$$

where the ϕ_i's are orthogonal eigenfunctions normalized to unity (orthonormal). We define the state vector as $|\psi\rangle$, which can be represented in terms of these eigenfunctions by a column vector:

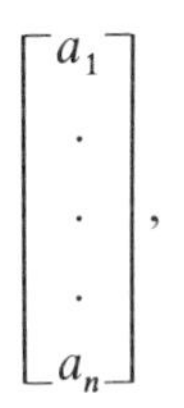

where $a_j = \int \phi_j^* \psi \, dV = \langle \phi_j | \psi \rangle$. The state vector which is equivalent to the complex conjugate wavefunction, $\langle \psi |$, can be expressed as a row vector:

$$[a_1^* \ldots a_n^*].$$

The difference between the wavefunction and the state vector is that the wavefunction describes the state for a particular coordinate choice whilst the state vector is independent of this choice.

Transformations, T, belonging to a good (i.e. valid) symmetry leave all expectation values unchanged, so that

$$\langle \psi | \psi \rangle = \langle T\psi | T\psi \rangle = \langle \psi | T^+ T | \psi \rangle,$$

where T^+ stands for the complex-conjugate transpose of T. Then $T^+ = T^{-1}$, i.e. T must be *unitary* for the expectation value to be unchanged. (For the case of time reversal, see Emmerson (1972).) Thus T may be constructed from some Hermitian operator D (i.e. an operator which has $D^+ = D$) by using:

$$T = \exp(iaD),$$

where a is numerical constant and D is known as the *generator* of the transformation. Observable quantities are represented by Hermitian operators so that we can sense here a connection between conserved observables and valid symmetries. This point is now pursued using four-momentum as an example of a conserved quantity. From its operator form $k_\rho = i\partial_\rho$ (given in section 1.1) a valid symmetry transformation can be constructed, namely

$$T = \exp(ia_p k_\rho),$$

where the a_ρ are four numerical constants. Let us consider the effect of this transformation on a plane wave of four-momentum P_μ (also given in section 1.1):

$$\psi(x_\mu) = A\exp(-iP_\mu x_\mu).$$

The effect is to replace the operator k_ρ by its eigenvalue P_ρ. Thus

$$\begin{aligned} T\psi(x_\mu) &= A\exp(ia_\rho P_\rho)\exp(-iP_\mu x_\mu) \\ &= A\exp[-iP_\mu(x_\mu - a_\mu)] \\ &= \psi(x_\mu - a_\mu). \end{aligned}$$

Therefore the transformations generated by the four-momentum operator are displacements in space-time. This result being true for plane waves is true for all states because any state can be written as a sum of plane waves. Therefore displacements in space-time belong to a valid symmetry of nature. This is clearly the case since we know that measurements of fundamental quantities like the rest mass of the electron don't depend on where or when the measurement is made. It is not difficult to show that the argument also works the other way round: nature is symmetric under displacements, therefore four-momentum is a conserved quantity. This argument generalizes to the statement that for every valid symmetry there exists a conserved quantity. Rotational symmetry is thus connected with the conservation of angular momentum. Rotations and displacements occur in space-time. There are also internal spaces; charge space, isospin space, etc., whose symmetries are also related to conserved quantities such as charge, isospin, etc.

5.2 Groups of transformations

A collection of transformations g form a group G if they satisfy these requirements:

(a) there is a composition law to define the effect of consecutive transformations, $g_1 g_2$,

(b) all products $g_1 g_2$ must also belong to the group,

(c) the inverse g^{-1} belongs to the group and also the identity $I = g^{-1}g$.

If all group elements commute ($g_2 g_1 = g_1 g_2$) then the group is called Abelian. Some groups are continuous like the displacement group or rotation group, in the sense that the displacement and the angle of rotation can vary continuously. Other groups like the parity group (involving reflection $\mathbf{r}\to-\mathbf{r}$) have a discrete number of elements. Transformations in space-time induce corresponding transformations on state vectors. For example, the coordinate transformation $\mathbf{r}\to g\mathbf{r}$ has the effect on a scalar wavefunction of transferring its value at $g^{-1}\mathbf{r}$ to the point $\mathbf{r}$ $[g(g^{-1}\mathbf{r})=\mathbf{r}]$. The wavefunction will now depend differently on the coordinates, so we call it $\psi'(r)$. However, there is still a connection with the old wavefunction $\psi(r)$, namely

$$\psi'(r)=\psi(g^{-1}r).$$

When the wavefunction is not a scalar (e.g. it is a vector or spinor) then the effect is more complicated. In terms of the basis eigenfunctions

$$\psi=a_j\phi_j \text{ (with implied summation over } j)$$

and

$$\begin{aligned}\psi'&=T(g)\psi=a_jT(g)\phi_j\\&=a_j\langle\phi_i|T(g)|\phi_j\rangle\phi_i\\&=a_jT_{ij}(g)\phi_i,\end{aligned}$$

where $T_{ij}=\langle\phi_i|T|\phi_j\rangle$. It is simple to show that these matrices $T_{ij}(g_1)$, $T_{ij}(g_2)$ . . . form a group with the same composition law as g_1, g_2 . . . . The matrices are said to provide a *representation* of the group G. Representations are called irreducible if the effect of any element of the group on any of the eigenfunctions ϕ_j produces another of these eigenfunctions or a linear superposition of them in the same representation. Often the word representation is applied loosely to mean the basis states ϕ_i. This convention will be used here when the sense is clear from the context.

5.3 Lie groups

We have seen that good symmetries do not change the intensity $|\psi|^2$. The transformations in the space of state vectors are therefore rotations which leave unchanged the complex lengths of state vectors. Such continuous transformations belong to groups of the type known as Lie groups. Lie groups are continuous and the transformations are functions of a finite set of real parameters. An infinitesimal transformation in one parameter a_i of a Lie group takes the form:

$$T(\ldots \delta a_i \ldots) = 1 + i\delta a_i X_i + \cdots,$$

which is close to the identity. Any finite transformation of a Lie group can always be built up from these infinitesimal ones:

$$T(\ldots a_i \ldots) = \operatorname*{Lt}_{n\to\infty} [1 + i(a_i/n)X_i]^n$$
$$= \exp(ia_i X_i).$$

The quantities X_i are recognized immediately as the generators (section 5.1). If successive small transformations are made:

$$T(\ldots \delta a_i \ldots)T(\ldots \delta a_j \ldots) = 1 + i\delta a_i X_i + i\delta a_j X_j$$
$$-\delta a_i \delta a_j X_i X_j + \cdots$$

and

$$T(\ldots \delta a_i \ldots)T(\ldots \delta a_j \ldots)T^{-1}(\ldots \delta a_i \ldots)$$
$$\times T^{-1}(\ldots \delta a_j \ldots)$$
$$= 1 - \delta a_i \delta a_j [X_i, X_j].$$

According to the group-composition law this is identical to some other group transformation

$$T(\ldots \delta a_k \ldots) = 1 + i\delta a_k X_k.$$

Thus

$$[X_i, X_j] = -i(\delta a_k/\delta a_j \delta a_i)X_k = C_{ijk} X_k.$$

However, the generators, X_i, do not depend on the δa_i coefficients, so it follows that the C_{ijk} coefficients do not depend on the δa_i coefficients either. The C_{ijk} are called the structure constants and the above equation defines what is called the group algebra of the generators. Once the group algebra is defined the basic structure of the group is defined. The minimum number of commuting generators of a Lie group is called its *rank*.

We now discuss some important Lie groups, starting with space-time symmetries (displacement symmetry was touched on in section 5.1).

5.4 Rotation group

Rotations about the origin in space have the property that the scalar products $x_i y_i$ are invariant, i.e. lengths of real vectors and the angles between them are unchanged. Rotations form a real orthogonal group, which means that the transpose of a transformation is equal to its inverse (i.e. $\tilde{T}=T^{-1}$ and $\tilde{g}=g^{-1}$). In n dimensions the group is called $0(n)$. The further restriction that the determinant det T is $+1$ excludes reflections in the origin ($\mathbf{r}\rightarrow-\mathbf{r}$). This restricted group is the special orthogonal group in n dimensions, called $S0(n)$. The generators for infinitesimal rotations about the axes in *three* dimensions are:

$$J_1=\begin{bmatrix}0 & 0 & 0\\ 0 & 0 & -i\\ 0 & i & 0\end{bmatrix}\quad J_2=\begin{bmatrix}0 & 0 & i\\ 0 & 0 & 0\\ -i & 0 & 0\end{bmatrix}$$

$$\text{and } J_3=\begin{bmatrix}0 & -i & 0\\ i & 0 & 0\\ 0 & 0 & 0\end{bmatrix}.$$

An infinitesimal rotation operator is then

$$R_i(\delta\theta)=\exp(iJ_i\delta\theta)\simeq I_3+iJ_i\delta\theta.$$

I_3 is a 3×3 matrix with 1's along the diagonal and zeros elsewhere. A finite rotation can be built from infinitesimal rotations thus:

$$\begin{aligned}R_1(\theta)&=\operatorname*{Lt}_{m\rightarrow\infty}\,[I_3+(i\theta/m)J_1]^m\\ &=I_3+J_1^2[-\theta^2/2!+\theta^4/4!-\cdots]\\ &\quad+iJ_1[\theta-\theta^3/3!+\cdots]\end{aligned}$$

$$= \begin{bmatrix} 1 & 0 & 0 \\ 0 & \cos\theta & \sin\theta \\ 0 & -\sin\theta & \cos\theta \end{bmatrix}.$$

Thus the new primed coordinates of a point P resulting from rotation of the coordinates through an angle θ about $0x$ are

$$\begin{bmatrix} x' \\ y' \\ z' \end{bmatrix} = R_1(\theta) \begin{bmatrix} x \\ y \\ z \end{bmatrix}.$$

The group algebra is

$$[J_i, J_j] = i\varepsilon_{ijk} J_k,$$

where ε_{ijk} is $+1$ for cyclic permutations of ijk (123, 231, 312); ε_{ijk} is -1 for anticyclic permutations (321, . . .); and ε_{ijk} is 0 for cases where any pair of i, j, k are equal (112, . . .). It follows (but we shall not prove it here) that there are two simultaneous observables. These are J^2 and any one generator J_3, i.e. the total angular momentum and one of its components. States which form the basis of an irreducible representation are labelled by the values taken by J and J_3, e.g. a state $|j, j_3\rangle$. J^2 is an example of a Casimir invariant: these are functions of the generators which commute with all generators. A group of rank r has r Casimir invariants. The Casimir invariants and the r mutually commuting generators constitute the simultaneous observables of a Lie symmetry group. The ladder or stepping operators

$$J_\pm = J_1 \pm iJ_2$$

give the well-known results that

$$J_+|j, m\rangle = \sqrt{(j+m+1)(j-m)}\,|j, m+1\rangle$$
$$J_-|j, m+1\rangle = \sqrt{(j+m+1)(j-m)}\,|j, m\rangle.$$

Each representation of the rotation group or rather its basis states can be displayed as points on a line, for which examples are shown in Fig. 5.1. These are called weight diagrams. If a symmetry holds then we would expect elementary particles to

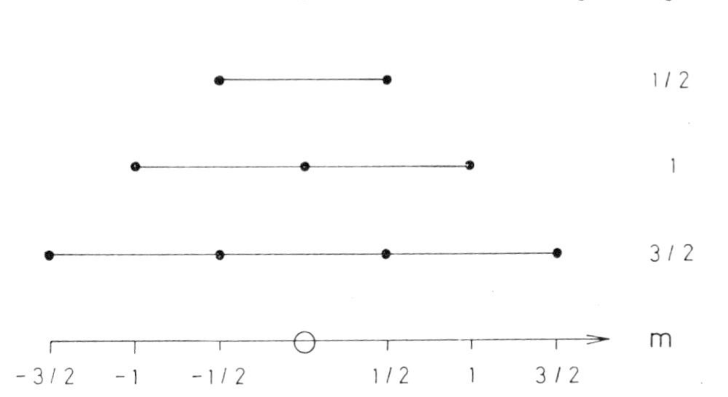

Fig. 5.1 Weight diagrams for representations of SO(3) with angular momenta 0, $\frac{1}{2}$, 1 and $\frac{3}{2}$.

belong to a particular representation and therefore to appear in multiplets which match the weight diagrams. Each irreducible representation is specified uniquely by the values of the Casimir invariants and each basis state of the representation is specified uniquely by the eigenvalues of the mutually commuting generators, i.e. by a unique point on the weight diagram. In a representation of angular momentum, j, the generators are represented by square $(2j+1)\times(2j+1)$ matrices with elements $\langle j, n|J_i|j, m\rangle$. These matrices obey the group algebra, of course.

Spherical harmonic functions form suitable orthonormal basis states for a representation of S0(3) in the case that j is *integral* (l). These are given by:

$$Y_l^m(\theta, \phi)=(-1)^{\frac{m+|m|}{2}}\sqrt{[(2l+1)(l-|m|)!/4\pi(l+|m|)!]}$$
$$P_l^{|m|}(\cos\theta)\exp(im\phi),$$

where $P_l^{|m|}$ is the associated Legendre polynomial, with total angular momentum l and third component m. The example used below will be:

$$Y_0^0=1/\sqrt{4\pi}, \quad Y_1^0=\sqrt{3/4\pi}\cos\theta$$

$$Y_1^1=-\sqrt{3/8\pi}\sin\theta\exp(i\phi) \text{ and}$$

$$Y_1^{-1}=\sqrt{3/8\pi}\sin\theta\exp(-i\phi).$$

Half-integral representations of S0(3) require the use of spinors, which in the non-relativistic case have two components

$$u_+ = \begin{bmatrix} 1 \\ 0 \end{bmatrix} \quad \text{and} \quad u_- = \begin{bmatrix} 0 \\ 1 \end{bmatrix}$$

for spin up and down respectively. The generators in this representation take the form:

$$J_i = \sigma_i/2,$$

where the Pauli matrices in a convenient form are

$$\sigma_1 = \begin{bmatrix} 0 & 1 \\ 1 & 0 \end{bmatrix}, \quad \sigma_2 = \begin{bmatrix} 0 & -i \\ i & 0 \end{bmatrix} \quad \text{and} \quad \sigma_3 = \begin{bmatrix} 1 & 0 \\ 0 & -1 \end{bmatrix}.$$

A rotation of α about an axis $\mathbf{n}$ with the polar angles θ, ϕ changes a spinor in the following way:

$$\begin{aligned} u \rightarrow R_n(\alpha)u &= \exp(i\alpha \mathbf{J} \cdot \mathbf{n})u \\ &= [I_2 \cos(\alpha/2) + i\boldsymbol{\sigma} \cdot \mathbf{n} \sin(\alpha/2)]u, \end{aligned}$$

where I_2 is the 2×2 unit matrix. A most unexpected result is that $R_n(2\pi)u = -u$, which shows that a rotation through 2π changes any spinor by a factor (-1). It requires a rotation of fully 4π to return a spinor to its initial value. Experiments carried out by Werner *et al.* (1975) confirmed this surprising quality of spinors. A beam of cold neutrons of 0·15 nm wavelength was directed at a single crystal cut in such a way as to leave three parallel but separated thin plates. Scattering from the first plate separated the beam into a direct component and a Bragg-scattered component. At the second plate the Bragg scattering of both components served to converge the beams again. At the third plate these coherent beams interfered. Their intensities were measured on emerging. Rotations of the neutron spin of one beam component were induced by applying a magnetic field over the path of that beam component only. For a spin rotation of 2π a change in the emerging beam intensity pattern was observed which matched what was expected for a phase reversal of the spin-rotated neutrons. When the spin rotation was doubled to 4π the intensity pattern returned to the field-off pattern.

This peculiarity of spinors can be explored further in group theory. We can write down the representation of any rotation in a spinor basis using the general form

$$R(\alpha)=\begin{bmatrix} a & -b^* \\ b & a^* \end{bmatrix}$$

with (aa^*+bb^*) equal to unity. It is simple to check that this matrix is both unitary and has determinant $+1$. The properties exactly match those of a rotation matrix in two dimensions where the dimensions are complex, i.e. they match the properties of a transformation belonging to the SU(2) group. The groups SU(2) and S0(3) therefore share the same algebra. The difference between the groups is that when the rotation angle in the two dimensions of SU(2) changes from 0 to 2π that in the three dimensions of S0(3) changes from 0 to 4π with consequences that are manifest in the behaviour of spinors under rotations.

From the simple fact that spinors exist we are led to the conclusion that nature has chosen SU(2) as the basic symmetry rather than S0(3). We lose none of the results of S0(3): they are simply supplemented. While we may live in a three-dimensional world the simplest representation of the rotation symmetry SU(2) is in the two dimensions inhabited by spinors. It is then very satisfying to find that *all* the fundamental material particles, quarks and leptons, have spin $\frac{1}{2}$.

The rules for adding angular momenta introduce us to another important property shared by other groups. When we add the states $|j, m\rangle$ and $|j', m'\rangle$ the resultant is a state $|J, M\rangle$ with J and M satisfying the rules:

$$|j-j'|\leqslant J\leqslant|j+j'|$$

and $M=m+m'$; J and M being integral (half-integral) if $(j+j')$ is integral (half-integral). In group-theoretical language the combination of two states $|j, m\rangle$ and $|j', m'\rangle$ is in a product representation, which can be decomposed as the sum of irreducible representations $|J, M\rangle$; i.e.

$$|j, m\rangle|j', m'\rangle=\sum_{J,M}(J, M|j, m\colon j', m')|JM\rangle$$

where the terms $(J, M|j, m; j' m')$ are numerical constants called Clebsch-Gordan coefficients. The following example shows how these coefficients are calculated. We add two spin-$\frac{1}{2}$ states $|\frac{1}{2}m_1\rangle_1$ and $|\frac{1}{2}m_2\rangle_2$; for $m_1 = m_2 = \frac{1}{2}$ the result is that

$$|\tfrac{1}{2}, \tfrac{1}{2}\rangle_1 |\tfrac{1}{2}, \tfrac{1}{2}\rangle_2 = |1, 1\rangle.$$

Applying the stepping operator J_- twice gives

$$|1, 0\rangle = \frac{1}{\sqrt{2}} [|\tfrac{1}{2}, -\tfrac{1}{2}\rangle_1 |\tfrac{1}{2}, \tfrac{1}{2}\rangle_2 + |\tfrac{1}{2}, +\tfrac{1}{2}\rangle_1 |\tfrac{1}{2}, -\tfrac{1}{2}\rangle_2],$$

and

$$|1, -1\rangle = |\tfrac{1}{2}, -\tfrac{1}{2}\rangle_1 |\tfrac{1}{2}, -\tfrac{1}{2}\rangle_2.$$

This completes a triplet of $J = 1$ states and there remains the $J = 0$ state $|0, 0\rangle$, which must be orthogonal to $|1, 0\rangle$ and must also be made from $m = \frac{1}{2}$ and $m = -\frac{1}{2}$ states. By inspection it is

$$|0, 0\rangle = \frac{1}{\sqrt{2}} [|\tfrac{1}{2}, -\tfrac{1}{2}\rangle_1 |\tfrac{1}{2}, +\tfrac{1}{2}\rangle_2 - |\tfrac{1}{2}, +\tfrac{1}{2}\rangle_1 |\tfrac{1}{2}, -\tfrac{1}{2}\rangle_2].$$

We see that, for example, the Clebsch-Gordan coefficient $(1, 0|\frac{1}{2}, -\frac{1}{2}; \frac{1}{2}, \frac{1}{2})$ has a value $1/\sqrt{2}$. The decomposition of the product of two doublet representatives into a singlet and triplet representation is written symbolically as:

representations $\quad 2 \otimes 2 = 1 \oplus 3$

angular momentum $\frac{1}{2} + \frac{1}{2} = \mathbf{0}$ or $\mathbf{1}$.

Addition of two or more angular momentum $\frac{1}{2}$ states can be used to produce any higher angular momentum state. The angular momentum $\frac{1}{2}$ representation is therefore called the *fundamental* representation. We expect that fundamental particles might belong to such representations and, as we have seen above, both leptons and quarks do possess spin $\frac{1}{2}$.

A representative set of Clebsch-Gordan coefficients for the rotation group S0(3) is given in Appendix C. These coefficients apply equally for SU(2), which shares the same group algebra as S0(3).

5.5 Lorentz and Poincaré groups

These are the groups of special relativity. Under Lorentz transformations the space-time interval (proper time) between events ($\mathbf{r}, t$) and ($\mathbf{r}', t'$) is invariant:

$$\tau^2 = (t - t')^2 - (\mathbf{r} - \mathbf{r}')^2.$$

The effect of boosting to a velocity β along the x_1-axis gives a transformation of space-time coordinates:

$$x \to x' = \begin{bmatrix} \cosh\mu & \sinh\mu & 0 & 0 \\ \sinh\mu & \cosh\mu & 0 & 0 \\ 0 & 0 & & \\ 0 & 0 & & I_2 \end{bmatrix} \begin{bmatrix} x_0 \\ x_1 \\ x_2 \\ x_3 \end{bmatrix}$$

where $\sinh\mu = \beta\gamma$ and $\cosh\mu = \gamma = (1-\beta^2)^{-\frac{1}{2}}$. This boost can be looked on as a rotation through an imaginary angle $i\mu$ in the $(x_1 - t)$ plane. The combination of rotations plus boosts forms the restricted Lorentz group S0(3, 1). The time component is distinguished because it appears with opposite sign in the expression for the proper time.

Displacement in space-time also leave τ^2 unchanged. The direct product of the displacement group and restricted Lorentz group gives the Poincaré group. This is the widest symmetry possessed by space-time consistent with the requirements of special relativity and the experimental results in particle physics. In other words, no experimental results are affected by any Poincaré transformation and, equivalently, particle interactions conserve (leave unchanged) the generators. The kinematic states of an elementary particle must therefore form irreducible representations of the Poincaré group: they cannot be drawn from several representations, otherwise a particle would exist in disconnected states. The choice of the observables for elementary particles (Casimir invariants plus commuting generators) is the following. The generators that commute are: p_μ the total four-momentum and $J_\mu p_\mu$, which for a single particle reduces to its helicity (λ). The Casimir invariants are p^2 and $p_0^2 J_i^2$, the latter evaluated in the centre-of-mass frame. For a single particle the Casimir invariants reduce to the mass squared, M^2, and $s(s+1)M^2$, where s is the spin. Mass and spin are thus the *inevitable* basic properties of particles in a universe where special relativity is valid. Helicity is the spin component along the direction of motion. For a photon the two helicity states are the states of circular polarization. $\lambda=1$ gives a right-handedly polarized photon: the electric- and magnetic-field vectors point perpendicular to the momentum and rotate at an angular frequency $E/\hbar$ about this axis.

E, $\mathbf{p}$ and λ depend on the reference frame but are conserved by all interactions. Helicity is the only truly measureable component of spin for a moving particle but at low enough velocities (the non-relativistic limit) the spin component, m, along an external axis becomes an alternative observable. Single-particle plane-wave states used in practical calculations are either chosen to be *helicity* states $|\mathbf{p}, \lambda\rangle$ or *canonical* states $|\mathbf{p}, m\rangle$. A convenient and conventional normalization is to have

$$\langle \mathbf{p}', \lambda' | \mathbf{p}, \lambda\rangle = 2E\delta(\lambda\lambda')\delta(\mathbf{p}-\mathbf{p}').$$

The discussion of particle states is complicated by the fact that detectors record the linear momenta of individual particles whilst the conserved quantities are the total four-momentum, total angular momentum and one component of angular momentum. There is some simplification if the frame of reference used is the overall centre-of-mass frame. Then a suitable choice for simultaneous observables is:

total CM four-momentum	$(E, 0)$
total angular momentum	J
total spin	S
total orbital angular momentum	L
component of J along external axis $(0z)$	M

where $\mathbf{J}=\mathbf{L}+\mathbf{S}$. With this choice the state vector is written $|JMLS\rangle$.

The experimentalist studying a two-body final state measures the polar angles (θ, ϕ) at which the particles emerge from the interaction and he may measure their polarizations m_1 and m_2 along $0z$ $(S_z=m_1+m_2)$. These measurements define a particle state written as $|\theta, \phi, S, S_z\rangle$ with normalization:

$$\langle \theta, \phi, s, s_z | \theta', \phi', s', s'_z \rangle = \delta(ss') \delta(s_z s'_z)\, \delta(\cos\theta - \cos\theta') \delta(\phi - \phi')$$

What is important in linking theory with experiment is therefore the relationship between the measured states $|\theta, \phi, S, S_z\rangle$ and states referred to conserved quantities $|\mathrm{JMLS}\rangle$ which are used in calculations. Explicitly

$$|\mathrm{JMLS}\rangle = \sum_{N, S_z} (J, M | L, N; S, S_z) \int \mathrm{d}\Omega\, Y_L^N(\theta, \phi) |\theta, \phi, S, S_z\rangle,$$

which for spinless particles reduces to the simple form

$$|JM\rangle = \int \mathrm{d}\Omega\, Y_J^M(\theta, \phi) |\theta, \phi\rangle.$$

5.6 Internal symmetries

These are symmetries acting directly on the particles yet not involving any displacement or rotation in space-time. Let us take strong isospin (see section 2.3) as an example of an internal symmetry. The strong interactions cannot distinguish a u-quark from a d-quark; or at the hadron level a proton (udd) from a neutron (udd). The u- and d-quarks can be classified as substates which point up and down in some space known as strong isospin space. Formally isospin is treated exactly like angular momentum; it is an SU(2) symmetry. The u-quark and d-quark are then two substates of one quark with strong isospin $\frac{1}{2}$: the u-quark has component $+\frac{1}{2}$ of the isospin along an arbitrary axis in isospin space while the d-quark has component $-\frac{1}{2}$. They can be written as $(\frac{1}{2}, +\frac{1}{2})$ and $(\frac{1}{2}, -\frac{1}{2})$. The fact that the strong isospin is conserved in strong interactions makes it possible to predict the relative rates of strong processes which involve different members of a strong isospin multiplet. An illustration is provided by the decays of $\Delta(1232)$ to a nucleon and a π-meson. One such decay is

$$\Delta^+ \rightarrow \mathrm{p} + \pi^0$$

$$(\tfrac{3}{2}, +\tfrac{1}{2}) \rightarrow (\tfrac{1}{2}, +\tfrac{1}{2}) + (1, 0)$$

where the numbers in brackets refer to the strong isospin. The projection of the product of the states $(\frac{1}{2}, +\frac{1}{2})$ and $(1, 0)$ on to the parent $(\frac{3}{2}, +\frac{1}{2})$ state is given by the Clebsch-Gordan coefficient $(\frac{3}{2},\frac{1}{2}|\frac{1}{2}, \frac{1}{2}; 1, 0)$, i.e. $\sqrt{\frac{2}{3}}$ from Appendix C. The Clebsch-Gordan coefficients of all the $\Delta(1232)$ baryon decays are:

$$
\begin{aligned}
&\Delta^{++} \rightarrow \mathrm{p}+\pi^{+}: && (\tfrac{3}{2}, \tfrac{3}{2}|\tfrac{1}{2}, \tfrac{1}{2}; 1, 1) && = 1 \\
&\Delta^{+} \rightarrow \mathrm{p}+\pi^{0}: && (\tfrac{3}{2}, \tfrac{1}{2}|\tfrac{1}{2}, \tfrac{1}{2}; 1, 0) && = \sqrt{\tfrac{2}{3}} \\
&\Delta^{+} \rightarrow \mathrm{n}+\pi^{+}: && (\tfrac{3}{2}, \tfrac{1}{2}|\tfrac{1}{2}, -\tfrac{1}{2}; 1, 1) && = \sqrt{\tfrac{1}{3}} \\
&\Delta^{0} \rightarrow \mathrm{p}+\pi^{-}: && (\tfrac{3}{2}, -\tfrac{1}{2}|\tfrac{1}{2}, \tfrac{1}{2}; 1, -1) && = \sqrt{\tfrac{1}{3}} \\
&\Delta^{0} \rightarrow \mathrm{n}+\pi^{0}: && (\tfrac{3}{2}, -\tfrac{1}{2}|\tfrac{1}{2}, -\tfrac{1}{2}; 1, 0) && = \sqrt{\tfrac{2}{3}} \\
&\Delta^{-} \rightarrow \mathrm{n}+\pi^{-}: && (\tfrac{3}{2}, -\tfrac{3}{2}|\tfrac{1}{2}, -\tfrac{1}{2}; 1, -1) && = 1.
\end{aligned}
$$

Then the *rates* of these decays are in the ratio 3:2:1:1:2:3. In weak and electromagnetic interactions strong isospin need not be conserved.

At this point we should remind ourselves that there is a fundamental distinction between *weak* and *strong* isospin even though they have identical mathematical structures. Weak isospin is evident in the pairing of quarks in the dominant weak decays. It is the more fundamental symmetry because it is the gauge symmetry underlying the weak force. On the other hand, strong isospin is an 'accidental' symmetry and does not lead to a gauge force (see Chapter 6).

The effect of symmetry transformations on particle states was shown above to be unitary, a feature which internal symmetries must share. The condition is satisfied by Lie groups because their representations are unitary or are equivalent to unitary representations. An internal space of the simplest sort has the familiar properties of the Hilbert space discussed in section 5.1. Its basis states are particle eigenstates and any single state can be expressed as a linear sum of such eigenstates with *complex* numerical coefficients, a_i. Linear transformations in this space

$$a_i \rightarrow b_i = A_{ij} a_j$$

which are subject to the unitary condition

$$\Sigma b_i b_i^* = \Sigma a_i a_i^*$$

form a Lie group known as the unitary group in n dimensions $U(n)$. These transformations are rotations that preserve the complex length $|\psi|$ of a vector. For any such rotation this definition leaves an overall phase ambiguity which can easily be removed by requiring that the determinant of the matrix A_{ij} is $+1$. When this requirement is imposed the restricted group is the special unitary group in n dimensions, SU(n). Basis states of the internal space form the simplest, or fundamental representation of the group SU(n).

Weak isospin has this character, being on SU(2) symmetry. Any group SU(n) has (n^2-1) generators and its fundamental representation has n basis states. We therefore expect the fundamental particles to group into n-fold fundamental multiplets if interactions satisfy an SU(n) symmetry. Otherwise these same particles could be constructed from fewer and hence more fundamental entities. The charge symmetry which was met in Chapter 4 is a U(1) symmetry (for one dimension U(1) is equivalent to SU(1)) so that all representations are singlets. What this implies is that elementary particles of a particular species (say electrons) all have the same charge; but it does not explain why there are simple numerical relationships between the lepton and quark charges. Whilst the gauge symmetry underlying quantum electrodynamics imposes only a trivial multiplet structure on the fundamental particles (all singlets) the symmetries underlying the other forces are more complex. Evidence for their existence is present in the multiplet structure of the leptons and quarks. This multiplet structure, once recognized, was used to suggest gauge symmetries for the strong force and for the unification of the electroweak forces. These are themes that will be developed in Chapters 6, 7 and 8. The symmetries of the standard model are all unitary symmetries but the other Lie symmetries may be important when it comes to extending our view (an example is supersymmetry, mentioned in Chapter 12).

5.7 Discrete symmetries

The discrete symmetries of interest have groups containing only two elements, one of which is the identity. In the case of parity the other transformation, P, is the reflection in the origin; $\mathbf{r} \rightarrow -\mathbf{r}$. P^2 is the identity and this 'closes' the group. Charge conjugation, C, exchanges particle for anti-particle. C^2 is the identity so that this group also only has two elements. Finally there is time reversal, T. The combination of transformations CPT is special because it is a good symmetry for any theory which is Lorentz invariant and only involves local interactions; requirements which any reasonable theory is expected to satisfy. If CPT holds, then a particle and its antiparticle must have *identical* masses and lifetimes.

Neither P nor T affect the proper time interval between events so it was long thought that they, like the restricted Lorentz transformations, would not affect the result of experiments. Lee and Yang (1956) suggested, and Wu *et al.* (1957) confirmed, that in weak decays this was not so. Wu *et al.* studied the β-decay of ^{60}Co at 0·01 K with its spin aligned in a magnetic field. Fig. 5.2 shows the current in the coil, the ^{60}Co spin alignment and the path of an electron from ^{60}Co decay. Under the parity transformation the current direction is unchanged, so the spin alignment of ^{60}Co is unaltered. On the

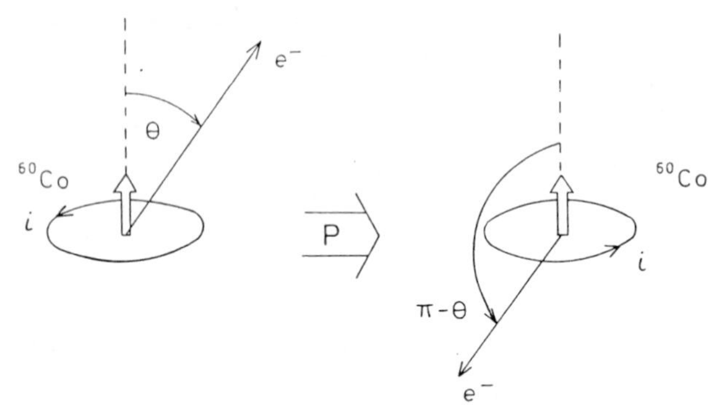

Fig. 5.2 The decay of a polarized ^{60}Co nucleus emitting an electron and the effect of the parity operation on this process.

other hand the electron direction is reversed. Mme Wu and her colleagues found that the angular distribution of the emitted electrons was proportional to $(1-\beta \cos \theta)$, where βc is the electron velocity. When the parity transformation is applied to this distribution it becomes $(1-\beta \cos(\pi-\theta))$, i.e. $(1+\beta \cos \theta)$, which contradicts the observations. Parity is therefore not a good symmetry because when it is applied to ^{60}Co-decay the transformed process is *not* seen in nature. Parity remains a good symmetry for the strong and electromagnetic processes including decays: the angular distributions from such decays of polarized parent states are symmetric about the equatorial plane with the angular distributions containing only even powers of $\cos \theta$. Notice that the parity operation reverses true vectors such as $\mathbf{r}$ and $\mathbf{p}$ but it leaves unchanged axial vectors formed by taking vector products, for example $\mathbf{r} \times \mathbf{p}$. Angular momentum $\mathbf{J}$ is therefore invariant under a parity transformation whilst the helicity $J_\mu p_\mu$ changes sign.

Individual particles possess intrinsic parities which can be consistently defined to be either $+1$ or -1. Referring to Appendix D we see that making a parity transformation on an electron spinor gives:

$$Pu(p, s)=\sqrt{E+m}\begin{bmatrix} \phi_s \\ \dfrac{-\boldsymbol{\sigma}\cdot\mathbf{p}}{(E+m)}\phi_s \end{bmatrix}=\gamma_0 u(p, s)$$

but for the positron spinor

$$Pv(p, s)=\sqrt{E+m}\begin{bmatrix} \dfrac{-\boldsymbol{\sigma}\cdot\mathbf{p}}{(E+m)}\chi_s \\ \chi_s \end{bmatrix}=-\gamma_0 v(p, s).$$

Thus, if the quarks and leptons are taken to have parity $+1$ their antiparticles have parity -1. Bosons have the same parity as their antiparticles; both have parity either $+1$ or -1. Orbital angular momentum eigenstates also have definite parity:

$$PY_L^M(\theta, \phi)= Y_L^M(\pi-\theta, \pi+\phi)=(-1)^L Y_L^M(\theta, \phi).$$

Any meson consists of a quark-antiquark pair moving with

some relative orbital angular momentum L. Hence the meson parity is $(-1)^{L+1}$. For the lightest mesons (π-, ρ-mesons, etc.) $L=0$, so they have negative parity. The baryons contain three quarks, so the lightest baryons all have positive parity. Parities have been determined for hadrons using either strong or electromagnetic processes. For example, the π^--meson parity was determined from the fact that capture at rest on deuterium produces a pair of neutrons. The deuteron has unit spin while the π-meson is spinless, and because the initial state has the π^--meson and deuteron at rest there is no relative orbital angular momentum. The total angular momentum for the initial state must be 1. Total angular momentum is conserved and hence the neutrons can have either zero relative orbital angular momentum ($L=0$) and spins parallel ($S=+1$), or $L=1$ and $S=0$. The former choice is ruled out by the Pauli exclusion principle. Thus the final state is a p state:

particles:	π^-	+	d (s state)	$\rightarrow$	n	+ n	(p state)
parity:	?		+	+	+	+	$-$

Parity is a multiplicative quantity hence $P(\pi^-)$ is negative. Under the parity transformation the electromagnetic field **A** reverses; consequently photons have *negative* parity.

The charge-conjugation operation changes a particle into its antiparticle. C parity, the analogue of parity, is conserved in strong and electromagnetic interactions but not in weak decays. The product CP turns out to be an almost perfect symmetry, being broken only at the level of 0·1 per cent and then only in the weak decays of neutral K-mesons. The general behaviour of CP is best illustrated using neutrinos which can only exist in the negative helicity state. Under a parity transformation helicity is reversed, which leads to an unobserved state. On the other hand, CP acting on a neutrino produces a positive helicity antineutrino, which is indeed the observed state for an antineutrino. C can be an observable only for neutral bosons. C acting on, for example, a π^--meson changes it into a π^+-meson. The photon has $C=-1$ because the photon couples to positive and negative charges in the same way. C parity is also a multiplicative quantum number, so that any neutral meson, such as the π^0, which decays to two photons must have positive C parity. Amongst the neutral mesons C parity is defined only for those made up from a quark and its own antiquark ($u\bar{u}$, for example). The C-parity operation affects both the spin and spatial wavefunctions:

$$C\chi_s(1, \bar{2})Y_L^M(\theta(1, \bar{2}), \phi(1, \bar{2})) = \chi_s(\bar{1}, 2)Y_L^M(\theta(\bar{1}, 2), \phi(\bar{1}, 2)),$$

where Y_L^M is the spatial wave function with $\theta(1, \bar{2})$ and $\phi(1, \bar{2})$ being the polar angles measured from the quark (1) to the antiquark ($\bar{2}$) and $\chi_s(1, \bar{2})$ is their spin wavefunction. The original state may be recovered from this by making the interchange $\bar{1} \leftrightarrow 2$. Thus the effect of C is the product of a factor η for the intrinsic C parity of a fermion-antifermion pair and the effect of label interchange on χ_s and Y_L^M. η is -1, the same as for parity because the same interchange is involved. Above it was seen that the spin triplet (singlet) is symmetric (antisymmetric) under label interchange. At the same time the spatial coordinates are reversed in Y_L^M thus changing it by a factor $(-1)^L$. The overall effect is a factor $(-1)^{L+S}$.

Summarizing the requirements of the quark model for *mesons* we have:

$$P = (-1)^{L+1} \text{ for all, and } C = (-1)^{L+S} \text{ for } u\bar{u},\ d\bar{d}, \text{ etc.,}$$

when the relative quark-antiquark orbital angular momentum is L and the total spin is S.

Time reversal is a good symmetry in strong, electromagnetic and most weak processes. In K^0-decay the breaking of CP must be accompanied by a compensating breaking of T in order that the product CPT remains a good symmetry. Time-reversal invariance implies that reactions and their inverses have identical amplitudes. This provides a way to determine the π-meson spin from a comparison of the rates of the reactions

$$p + p \rightleftharpoons \pi^+ + d.$$

If the beams and targets are unpolarized the cross-sections are (see Appendix A)

$$d\sigma_1 = \frac{1}{(2S_p+1)^2} \sum_{\text{spin}} |T|^2 \, dQ_2(\pi, d)$$

$$d\sigma_2 = \frac{1}{2(2S_\pi+1)(2S_d+1)} \sum_{\text{spin}} |T|^2 \, dQ_2(pp)$$

where the factor 2 is needed to account for antisymmetrization of the wavefunctions of the two emerging protons. At low energies the interaction is s-wave so that the amplitude T contains no kinematic factors. Then using the expression given for dQ_2 in the appendix

$$(d\sigma_1/d\Omega^*) = [3(2S_\pi+1)/2]\,(p^*_\pi/P^*_p)^2 (d\sigma_2/d\Omega^*).$$

At identical CM energies the total cross-sections for the reactions are respectively $0{\cdot}18 \pm 0{\cdot}06$ mb and $3{\cdot}1 \pm 0{\cdot}3$ mb with $(p^*_\pi/P^*_p) = 0{\cdot}201$; thus s_π (integral) is zero (Cartwright (1953)).

A last discrete symmetry group of interest is that of particle permutation. If two identical fermions are permuted then the overall state changes sign, whilst if two identical bosons are permuted the state is unchanged. These are the statements underlying Fermi-Dirac and Bose-Einstein statistics; a fermion state can contain at most one fermion whereas a boson state can contain any number of bosons. It is very important to realize that permutation applies simultaneously to the whole wavefunction: it applies to the product of the spatial part, the spin part and any internal parts.

5.8 Summary on conservation laws

Table 5.1 summarizes the conservation laws so far discussed as well as anticipating some to be met later in Chapters 6–8.

Table 5.1

Symmetry	*Electroweak force* *Weak sector*	*E.M. sector*	*Strong force*
Space-time symmetries			
4-momentum, total angular momentum and component	√	√	√
C,P,T	×	√	√
CPT	√	√	√
Internal gauge symmetries			
Charge	√	√	√
Colour	√	√	√
Other internal symmetries			
Quark flavour	×	√	√
Strong isospin	×	×	√
Fermion conservation			
Baryon number e, μ, τ lepton number	√	√	√

Chapter 6
Colour and QCD

Careful examination of the wavefunctions of baryons shows that, thus far, they are symmetric rather than antisymmetric under the interchange of the quarks, although the quarks are fermions. It requires the introduction of a new quantum number, known as colour, to achieve the necessary antisymmetrization. With the new quantum number comes a new 'colour' symmetry, which is recognized to be the symmetry underlying the quantum field theory of the strong interaction (quantum chromodynamics). We shall trace these developments in this chapter. The quanta of the colour field are called gluons and, like photons, they are massless. Gluons carry 'colour' charge and therefore interact with each other, unlike the photons. This feature has the consequence that the colour force increases in strength with separation; just the opposite of the behaviour of the electromagnetic force. The strong force possesses an energy scale and a complementary length scale which have consequences for the size of hadrons and for the existence of flavour symmetries. Together the quarks, antiquarks and gluons constitute the totality of strongly interacting fundamental particles. Collectively they are called *partons*.

6.1 SU(3) The colour symmetry

The quark content of the lowest mass hadrons is shown in Fig. 2.3. Inside these hadrons the quarks are in states of zero relative orbital angular momentum (*S* states), hence the hadron spin is just the vector sum of the quark spins. The Δ^{++} structure (uuu) presents an immediate problem. This baryon has intrinsic angular momentum 3/2, which means that the spins of the three u quarks are aligned parallel. The spatial and spin states of the Δ^{++}-baryon are therefore symmetric under interchange of any pair of its component quarks. This situation is untenable because quarks are fermions and their wavefunctions ought to be antisymmetric under their mutual interchange. Put another way, the Pauli principle is violated by having three quarks in the same quantum state. In 1964 this basic difficulty of the quark model was resolved when Greenberg proposed that a hitherto unrecognized symmetry was involved, called colour. The quark wavefunction describing a hadron's internal structure must be extended to four components:

$$\psi = (\text{space term}) \times (\text{spin term}) \times (\text{flavour term}) \times (\text{colour term}).$$

Overall antisymmetry of the Δ^{++} total wavefunction is restored if the colour term is antisymmetric under quark interchange. Three different quark-quark interchanges are possible within a baryon so that colour must be a threefold symmetry; SU(3) was the simple choice made by Greenberg. Quarks thus carry a colour quantum number which has three distinct possible values (red, yellow, blue) in addition to spin and to flavour (up, down, strange, . . .). As fundamental particles the quarks provide a fundamental three-dimensional representation of the colour group SU(3).

What are conspicuously absent in nature are any corresponding coloured hadron multiplets; for example, mesons made from red quarks, others made from blue quarks and yet others made from yellow quarks. We could then expect to observe, for example, nine π^{+}-mesons which might differ in mass if the quarks' mass depended on its colour. Instead, the hadron spectrum is fully explained without recourse to colour. The most economic conclusion is that quarks differing only in colour have identical masses and that all the hadrons are colour singlets. Why, then, are hadrons colour singlets? The answer, as we shall see later, lies in the group structure of the SU(3) symmetry.

Colour is an extremely good symmetry and so, paradoxically, direct demonstrations of colour effects are hard to find. However, all the electroweak field particles (γ, $W^{\pm}$, Z^{0}) are colourless and can only couple to a quark-antiquark pair of matching colour. This in turn requires that the rates of certain electroweak processes are enhanced, if colour exists, over their rates in the absence of colour. According to the quark model the reaction

$$e^{+}+e^{-}\rightarrow \text{hadrons}$$

proceeds by the multi-step process shown in Fig. 6.1(a). The

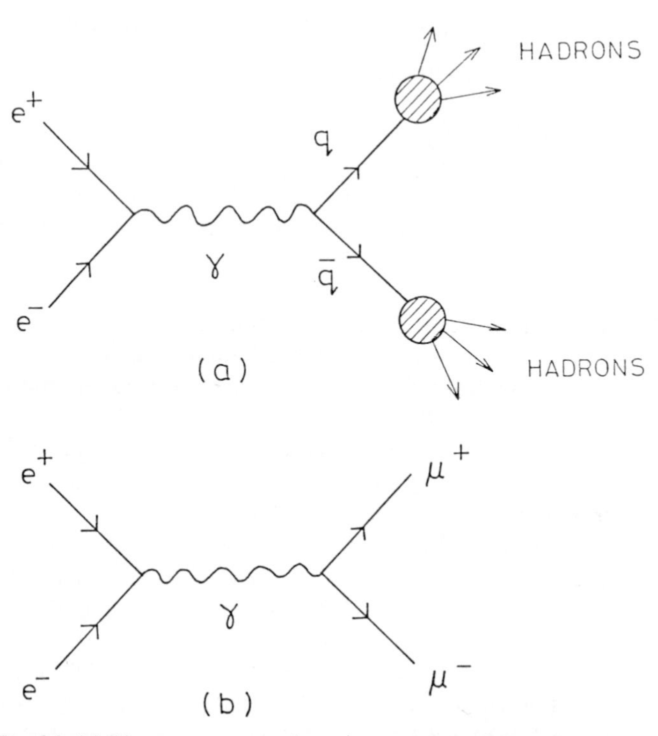

Fig. 6.1 (a) The two-step reaction $e^{+}+e^{-}\rightarrow\bar{q}+q$ followed by hadronization. (b) The dominant Feynman diagram for $e^{+}+e^{-}\rightarrow \mu^{+}+\mu^{-}$.

annihilation gives a virtual photon that converts to a quark-antiquark pair and subsequently the quarks fragment into hadrons. This last step has unit probability because only hadrons are observed, never free quarks. The rate is exactly that for

$$e^+ + e^- \rightarrow q + \bar{q}. \quad [6.1]$$

The resemblance to another process involving point-like fermions

$$e^+ + e^- \rightarrow \mu^+ + \mu^- \quad [6.2]$$

shown in Fig. 6.1(b), was noted in Chapter 4. The rates of these two reactions [6.1] and [6.2] will differ by a factor equal to the square of the quark charge (Q^2). There will be a further factor of 3 if the quark colour is a good quantum number because then there are three distinguishable quark-antiquark colour combinations. The ratio of the rates for the processes [6.1] and [6.2] is:

$$R = \frac{\sigma(e^+ + e^- \rightarrow \text{hadrons})}{\sigma(e^+ + e^- \rightarrow \mu^+ + \mu^-)} = 3\sum_f Q_f^2,$$

where the sum is carried out over the active quark flavours. If the quark of flavour f has mass m_f then the CM energy will need to exceed $2m_f c^2$ before this quark species can be included in the sum as an active flavour. Above 10 GeV CM energy the u, d, s, c and b flavours are all produced and the prediction is that

$$R = 3\left(\frac{2^2}{3} + \frac{1^2}{3} + \frac{1^2}{3} + \frac{2^2}{3} + \frac{1^2}{3}\right) = \frac{11}{3} = 3{\cdot}66.$$

Measurements made at the PETRA electron positron collider at DESY at 40 GeV CM energy (Behrend (1984)) give

$$R = 3{\cdot}9 \pm 0{\cdot}3,$$

which agrees well with the three-colour hypothesis.

Parallel arguments can be made using the decay rates of the neutral pseudoscalar mesons (spin parity 0^-) to photons: $\pi^0 \rightarrow 2\gamma$, $\eta^0 \rightarrow 2\gamma$, $\eta^{0\prime} \rightarrow 2\gamma$. The rate of decay, Γ, depends on the probability that the component quark and antiquark overlap and annihilate:

$$\Gamma \propto |s|^2,$$

where s is the sum of contributions from each quantum mechanically distinguishable quark eigenstate. With three quark colours Γ is boosted by a factor 9. A width of about 9 eV is expected for the decay of a π^0 made from coloured quarks, compared to around 1 eV in the absence of colour. Experimentally, Atherton *et al.* (1985) measured a width of $7{\cdot}43 \pm 0{\cdot}3$ eV.

6.2 The group properties of SU(3)

The properties of SU(3) have similarities with those of SU(2). Each member of a representation needs two labels rather than one: the labels are I_3^c, the third component of colour isospin, and Y^c the colour hypercharge. Fig. 6.2 shows the weight diagrams for the two inequivalent fundamental representations, 3 and $\bar{3}$. These representations accommodate respectively the quarks and the antiquarks. (We write r for a red quark, $\bar{\mathrm{r}}$ for a red antiquark and so on.) The quarks and antiquarks are then distinguishable and have opposite eigenvalues: $I_3^c(\bar{\mathrm{q}}) = -I_3^c(\mathrm{q})$ and $Y^c(\bar{\mathrm{q}}) = -Y^c(\mathrm{q})$. We can write the three quark basis states as

$$\mathrm{r} = \begin{bmatrix} 1 \\ 0 \\ 0 \end{bmatrix}, \mathrm{y} = \begin{bmatrix} 0 \\ 1 \\ 0 \end{bmatrix} \text{ and } \mathrm{b} = \begin{bmatrix} 0 \\ 0 \\ 1 \end{bmatrix}.$$

There are eight $(3^2 - 1)$ generators of SU(3) and these are written

$$F_i = \lambda_i/2.$$

In the fundamental representation with the quarks as basis states:

$$\lambda_1 = \begin{bmatrix} 0 & 1 & 0 \\ 1 & 0 & 0 \\ 0 & 0 & 0 \end{bmatrix}, \lambda_2 = \begin{bmatrix} 0 & -i & 0 \\ i & 0 & 0 \\ 0 & 0 & 0 \end{bmatrix},$$

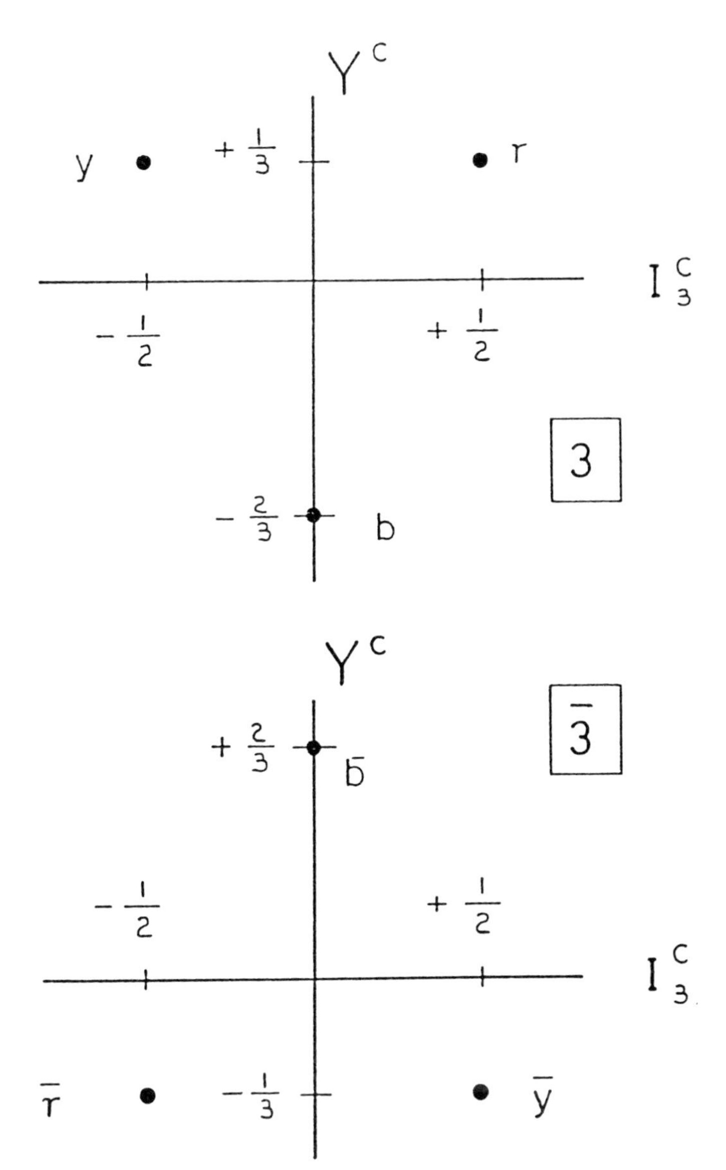

Fig. 6.2 The weight diagrams for the two inequivalent fundamental representations of SU(3) of colour, 3 and $\bar{3}$.

$$\lambda_3 = \begin{bmatrix} 1 & 0 & 0 \\ 0 & -1 & 0 \\ 0 & 0 & 0 \end{bmatrix}, \lambda_4 = \begin{bmatrix} 0 & 0 & 1 \\ 0 & 0 & 0 \\ 1 & 0 & 0 \end{bmatrix},$$

$$\lambda_5 = \begin{bmatrix} 0 & 0 & -i \\ 0 & 0 & 0 \\ i & 0 & 0 \end{bmatrix}, \lambda_6 = \begin{bmatrix} 0 & 0 & 0 \\ 0 & 0 & 1 \\ 0 & 1 & 0 \end{bmatrix},$$

$$\lambda_7 = \begin{bmatrix} 0 & 0 & 0 \\ 0 & 0 & -i \\ 0 & i & 0 \end{bmatrix}, \lambda_8 = \frac{1}{\sqrt{3}} \begin{bmatrix} 1 & 0 & 0 \\ 0 & 1 & 0 \\ 0 & 0 & -2 \end{bmatrix}.$$

The group algebra is:

$$[F_i, F_j] = i f_{ijk} F_k.$$

Its non-zero structure constants are $f_{123} = 1$, $f_{147} = f_{246} = f_{257} = f_{345} = f_{516} = f_{637} = \frac{1}{2}$, $f_{458} = f_{678} = \sqrt{3}/2$: structure constants obtained by cyclic permutations of the subscripts have the same values, while anticyclic permutations have the sign reversed. A rotation of θ about an axis $\mathbf{n}$ in the internal colour space produces the following transformation of the quark colour wavefunctions χ:

$$U\chi = \exp(i\theta F_a n_a)\chi,$$

in which the colour index a runs from 1 to 8.

There are four simultaneous colour observables, which can be chosen as the two diagonal generators F_3 and F_8 and two Casimir invariants, one of which is F^2. Weight diagrams of representations of colour are therefore two-dimensional. The standard choice of coordinates on the weight diagrams are the parameters mentioned above:

$$I_3^c = F_3 \text{ (third component of colour isospin)}$$

and

$$Y^c = 2F_8/\sqrt{3} \text{ (colour hypercharge).}$$

Higher-order representations are built from products of 3 and $\bar{3}$. The weight diagram for $3 \otimes \bar{3}$ is obtained by lifting the $\bar{3}$ weight diagram and centring it at y, r and b in turn. The resulting nine combinations are shown in Fig. 6.3. One of the combinations made up from the overlapping $\bar{r}r$, $\bar{y}y$ and $\bar{b}b$ states with $I_3^c = Y^c = 0$ is

$$\frac{1}{\sqrt{3}}[r\bar{r} + y\bar{y} + b\bar{b}].$$

Every generator annihilates this state so it must be a colour singlet; it describes the colour content of the *mesons*. The full reduction of this product representation is:

$$3 \otimes \bar{3} = 8 \oplus 1.$$

The generators themselves belong to the octet representation. This contains the six peripheral states forming a hexagon in

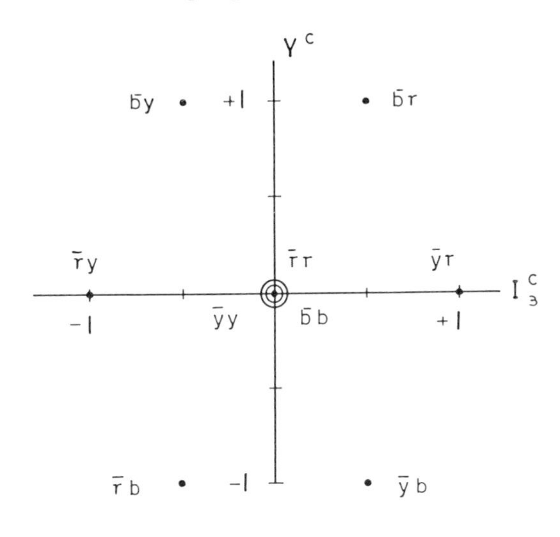

Fig. 6.3 The weight diagram for the product representation $\bar{3} \otimes 3$ of SU(3) colour.

Fig. 6.3 plus the two remaining combinations of $r\bar{r}$, $y\bar{y}$ and $b\bar{b}$ that are orthogonal to $(1/\sqrt{3})[r\bar{r}+y\bar{y}+b\bar{b}]$. It is worth noting that the representations 8 and $\bar{8}$ are identical. The 8 is therefore called the self-adjoint representation. The weight diagrams of all SU(3) representations have certain common features: they are either equilateral triangles, or the hexagons formed by cutting off the corners of an equilateral triangle symmetrically. In what follows we shall need the expectation value of the Casimir invariant F^2 for the fundamental representation. This can be calculated using any of the quarks, e.g. for the red quark state:

$$\langle F^2 \rangle = [100] F^2 \begin{bmatrix} 1 \\ 0 \\ 0 \end{bmatrix}$$

$$= [100] \begin{bmatrix} \frac{4}{3} & 0 & 0 \\ 0 & \frac{4}{3} & 0 \\ 0 & 0 & \frac{4}{3} \end{bmatrix} \begin{bmatrix} 1 \\ 0 \\ 0 \end{bmatrix} = \frac{4}{3}.$$

Below we tabulate the expectation values $\langle F^2 \rangle$ for several simple representations.

Table 6.1

Representation	1	3 or $\bar{3}$	6	8
F^2	0	$\frac{4}{3}$	$\frac{10}{3}$	3

We now return to the question of hadron colour and see how this relates to the quark content of the hadrons. Taking the simplest products of quarks and antiquarks their product representation can be shown to resolve as follows:

$q\bar{q}$ $\quad 3\otimes\bar{3}=1\oplus 8$

qq $\quad 3\otimes 3=\bar{3}\oplus 6$

qqq $\quad 3\otimes 3\otimes 3=1\oplus 8\oplus 8\oplus 10$

$qq\bar{q}$ $\quad 3\otimes 3\otimes\bar{3}=3\oplus 3\oplus\bar{6}\oplus 15.$

What this resolution reveals is that *only* the combinations q$\bar{\text{q}}$ (mesons) and qqq (baryons) are capable of giving colour singlets. Furthermore, of the qqq representations, only the singlet has the necessary antisymmetric behaviour under quark-quark interchange. More complicated arrangements such as qqq(q$\bar{\text{q}}$) are possible: they may exist as excited states which return to the form qqq by q$\bar{\text{q}}$ annihilation. Correspondingly q$\bar{\text{q}}$ annihilation is a mechanism by which mesons may decay.

6.3 The colour force

Electrodynamics provides the pattern on which to model the gauge theory of the colour force. The starting point is to impose invariance on the Dirac equation for quarks

$$(i\gamma_\mu \partial_\mu - m)\psi = 0$$

under an infinitesimal local colour transformation

$$\psi \to \exp[ig_s F_a n_a \theta(x)]\psi = (1 + ig_s F_a n_a \theta(x))\psi,$$

where $\theta(x)$ is the rotation angle in colour space at a point x in space-time and g_s is the strength of the colour 'charge'. Under such transformations the simple derivative is not invariant, and following the approach used in Chapter 4 for electromagnetism we replace it by the covariant derivative:

$$D_\mu = \partial_\mu + ig_s F_a G_{a\mu}(x).$$

The gauge field $G_{a\mu}$ is a vector in space-time and carries colour. Its product with F_a must be a colour singlet, so it is natural to assign G like F to the self-adjoint (8) representation. G must transform in a particular way under local colour transformations if the covariant derivative is to remain invariant under such transformations:

$$G_{a\mu}(x) \to G_{a\mu}(x) - n_a \partial_\mu \theta(x) - g_s f_{abc} n_b G_{c\mu}(x)\theta(x).$$

Lastly the energy of the gauge field takes the form

$$G_{a\mu\nu} G_{a\nu\mu}$$

where

$$G_{a\mu\nu}=\partial_\mu G_{a\nu}-\partial_\nu G_{a\mu}-g_s f_{abc}G_{b\mu}G_{c\nu}.$$

There are strong similarities with electromagnetism (as the reader should verify by returning to Chapter 4) and also significant differences. First there is a gauge field G which, like the electromagnetic field, is a vector in space, but unlike the latter has eight colour components. There are correspondingly eight quanta of the colour field, called gluons. Looking back at Fig. 6.3 we remark that they match the states $\bar{\mathrm{b}}\mathrm{y}$, $\bar{\mathrm{b}}\mathrm{r}$, etc. Again by analogy with photons these gluons couple to the colour charge of the quarks with a minimal colour coupling and strength g_s. The analogue of the fine-structure constant is $\alpha_s=g_s^2/4\pi$. Measured at energies of order 1 GeV α_s is about unity, reflecting the much greater strength of the strong interaction. Finally it can be inferred from the requirement of gauge invariance of the gluon field energy density that the gluons are massless. What is very different from the case of electromagnetism is that the energy density contains terms cubic and quartic in the field. This feature has its origin in the non-Abelian nature of the colour group SU(3), in contrast with the Abelian charge group U(1). When the implications of such terms are worked out it transpires that three or four gluons may interact at a point. Then whereas electromagnetic fields obey the simple superposition principle colour fields interact to give non-linear effects.

The quantized form of the colour field theory is known as quantum chromodynamics (QCD). Fig. 6.4 shows the possible vertices. When a quark-gluon-quark coupling vertex (Fig. 6.4(a)) appears in the amplitude for a reaction it contributes a factor

$$ig_s\gamma_\mu[F_a]_{bc},$$

where $[F_a]_{bc}$ is the element in the b'th row and c'th column of the matrix F_a in the 3 representation. By analogy with the electromagnetic Coulomb force the single-gluon exchange gives rise to a force of the form

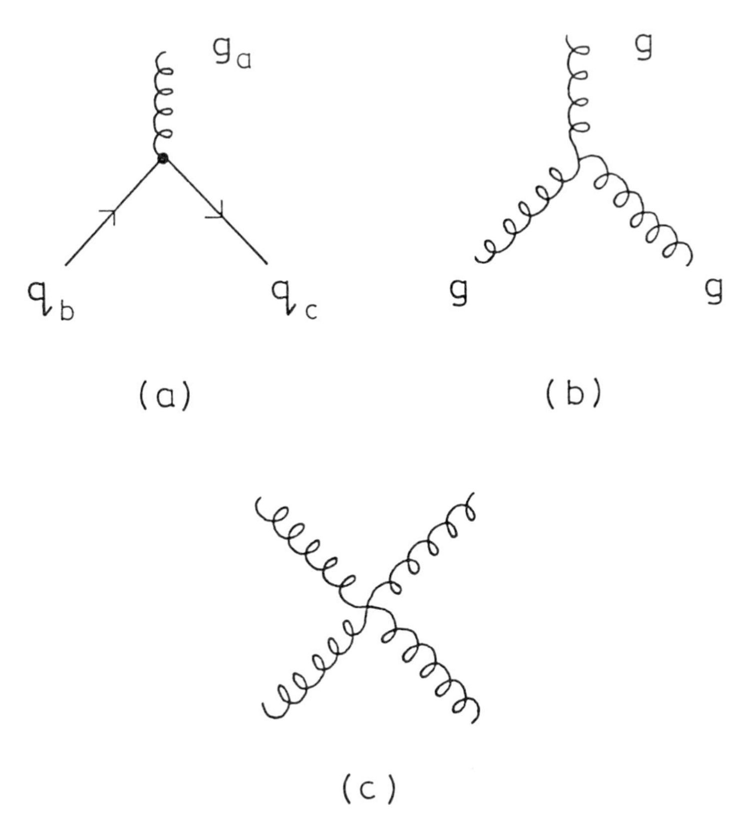

Fig. 6.4 (a) The quark-quark-gluon vertex, where the subscripts indicate the colour indices. (b) The three-gluon vertex. (c) The four-gluon vertex.

$$\mathbf{F}_{g} = \frac{1}{r^2}\,\mathbf{F}_1 \cdot \mathbf{F}_2,$$

where $\mathbf{F}_1$ and $\mathbf{F}_2$ are vectors whose components are the colour generators. $\mathbf{F}_1$ refers to the first of the two interacting quarks or antiquarks and $\mathbf{F}_2$ to the second. $\mathbf{F}_1 \cdot \mathbf{F}_2$ replaces the product of charges in a theory based on an Abelian symmetry.

$$\begin{aligned}\mathbf{F}_1 \cdot \mathbf{F}_2 &= \tfrac{1}{2}[(\mathbf{F}_1+\mathbf{F}_2)^2 - F_1^2 - F_2^2]\\ &= \tfrac{1}{2}(F^2 - F_1^2 - F_2^2),\end{aligned}$$

where F^2 is the value of the Casimir invariant in the colour representation to which the two-quark (antiquark) system belongs. For qq, or q$\bar{q}$ or $\bar{q}\bar{q}$ we have (from Table 6.1):

$$\begin{aligned} \mathbf{F}_1 \cdot \mathbf{F}_2 &= F^2/2 - \tfrac{4}{3} \\ &= -\tfrac{4}{3} \qquad q\bar{q} \text{ in } 1 \\ &= -\tfrac{2}{3} \qquad qq \text{ in } \bar{3} \\ &= -\tfrac{2}{3} \qquad \bar{q}\bar{q} \text{ in } 3 \end{aligned}$$

Higher representations have $\mathbf{F}_1 \cdot \mathbf{F}_2$ positive so that an attraction is only likely for the representations 1, 3 and $\bar{3}$. In the case of the baryons, a qq pair in the $\bar{3}$ representations interacts with a q in the 3-representation so that

$$\mathbf{F}_1 \cdot \mathbf{F}_2 = -\tfrac{4}{3} \qquad (qq)q \text{ in } 1$$

and once more the force is likely to be attractive. Binding of both q$\bar{q}$ and qqq systems is thus possible with a non-Abelian symmetry, which contrasts vividly with the case of electric charge.

6.4 Quantum chromodynamics

The quantum field theory of the colour force is known as quantum chromodynamics (QCD). In principle the amplitude for any two-body to n-body process can be obtained by drawing the Feynman diagrams of importance and adding up their contributions to the amplitude. There is a set of rules, similar to those for QED, which gives the factors associated with external lines, internal lines and vertices. One important difference lies in the existence of the new types of vertices (Fig. 6.4(b) and 6.4(c)), which are only possible for a non-Abelian symmetry. Another significant difference is the appearance of the colour factors noted for Fig. 6.4(a). However, the simple idea of using a perturbation expansion to calculate an amplitude can break down for QCD. At the nuclear scale (10^{-15} m) the coupling constant g_s must be of order unity because cross-sections for hadronic processes at ~1 GeV energy are ten thousand times larger than for electromagnetic processes. Each vertex added to a diagram contributes a further factor g_s in the amplitude so that higher-order diagrams are just as important as simpler diagrams. The perturbation expansion fails, therefore, at ranges $\sim 10^{-15}$ m. On the other hand, deep (i.e. involving large momentum transfer) inelastic lepton scattering from nucleons has the appearance of scattering from free point-like constituents. (This result was introduced in Chapter 2 and will be further developed in Chapter 9.) To summarize, the quarks in nucleons behave as if well bound for momentum transfers that are small ($-q^2 \ll 1$ GeV^2) while for large momentum transfers ($-q^2 \gg 1$ GeV^2) they behave as if free. QCD needs somehow to reproduce these features.

Politzer (1973) and Gross and Wilczek (1973) proved that in the limit of large four-momentum transfers the coupling strength in QCD tends to zero. This is called asymptotic freedom because quarks become asymptotically free at high momentum transfers q^2 or, equivalently, at small separations. Asymptotic freedom is a unique property of non-Abelian gauge theories. The result can be made plausible by making arguments similar to those already used to explain the effects of vacuum polarization in QED. Fig. 6.5 shows loop diagrams which lead to colour polarization of the vacuum. We have met such loops in connection with QED and found that it was necessary to absorb their divergent contributions by *renormalizing* the coupling strength. In QCD there is an important difference that whereas the quark loops result in *shielding* of the colour charge the gluon loops carry colour and *enhance* the colour charge. The resulting renormalized coupling strength has a dependence on the four-momentum transfer square (q^2) of the form

$$\alpha_s(q^2) = \alpha_s(\mu^2)/\{1 + B\alpha_s(\mu^2)\ln(|q^2|/\mu^2)\},$$

where $\alpha_s(\mu^2)$ is evaluated at some reference four-momentum and where B contains terms from vacuum polarization and other effects. B has the form

$$B = [11 - 2N_F/3]/4\pi,$$

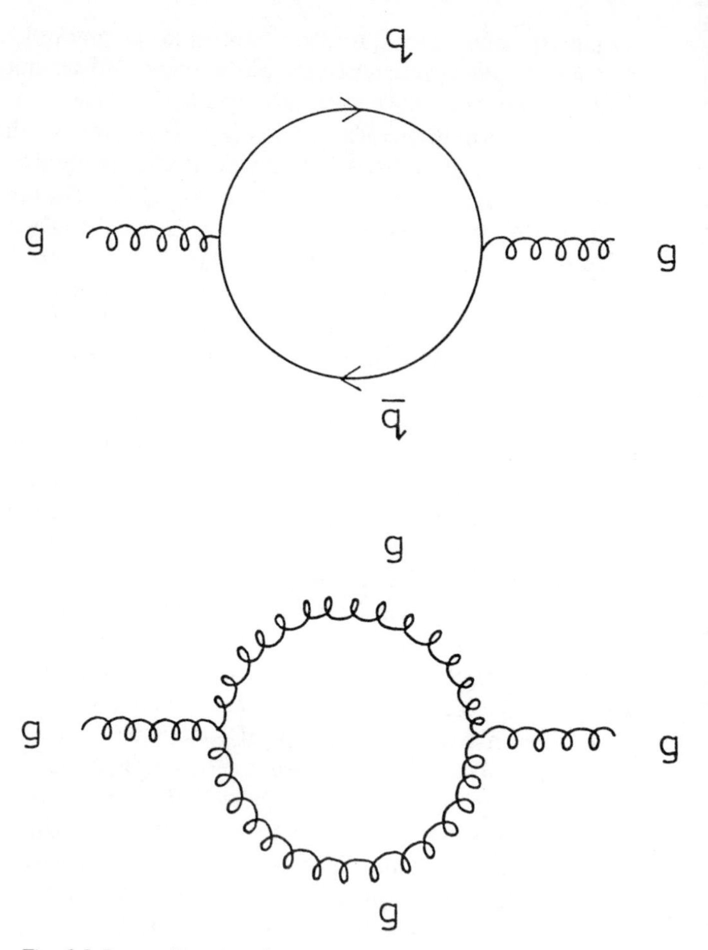

Fig. 6.5 Loop diagrams for colour polarization of the vacuum.

where N_F is the number of quark flavours. With N_F set to six, as appears to be the case, B is positive and

$$\alpha_s(q^2) \rightarrow 0 \text{ as } -q^2 \rightarrow \infty,$$

which leads to asymptotic freedom. Notice that if there were very many species (flavours) of quarks the quark-loop contributions would dominate over the gluon loops; B would be negative and then the coupling strength α_s would increase with q^2. In QED there are no photon loops so that α does increase with q^2.

The first direct observations of parton-parton scattering have been made in experiments at the CERN SPS Collider by Arnison *et al.* (1983), and Banner *et al.* (1982). Within this machine, which is basically a proton synchrotron, protons and antiproton bunches circulate in opposite senses. These beams are simultaneously accelerated to 315 GeV (currently) and then made to collide at the centre of two giant detectors (UA1 and UA2) spaced 60° apart around the 2 km diameter accelerator ring. Collimated 'jets' of hadrons are clearly seen emerging from a proportion of the interactions and are principally the end products of an 'elastic' scattering process:

$$\text{parton}_1 + \text{parton}_2 \rightarrow \text{parton}_3 + \text{parton}_4$$

The partons may be quarks, antiquarks or gluons; the first parton coming from the proton and the second from the antiproton. Subsequently partons 3 and 4 turn into jets of hadrons by a process called 'fragmentation', with each jet of hadrons retaining the four-momentum of its parent parton. More will be said about fragmentation in section 11.2. The modulus of the four-momentum transfer squared between the colliding partons in the data is greater than 1600 GeV^2, so that the coupling strength has become quite small. Consequently the dominant contribution to scattering amplitudes should come from single-gluon exchange between the partons. This immediately implies that the scattering angular distribution should resemble that for Rutherford scattering, because both processes are mediated by the exchange of a single massless vector meson:

$$d\sigma/d\cos\theta^* \propto q^{-4} \propto (1-\cos\theta^*)^{-2},$$

where θ^* is the CM scattering angle. Fig. 6.6 compares the measurement of Arnison *et al.* (1984) with this prediction for the angular distribution using vector gluons. The agreement is excellent.

Other direct evidence of the existence of the gluon comes from the observation of processes in which three jets emerge.

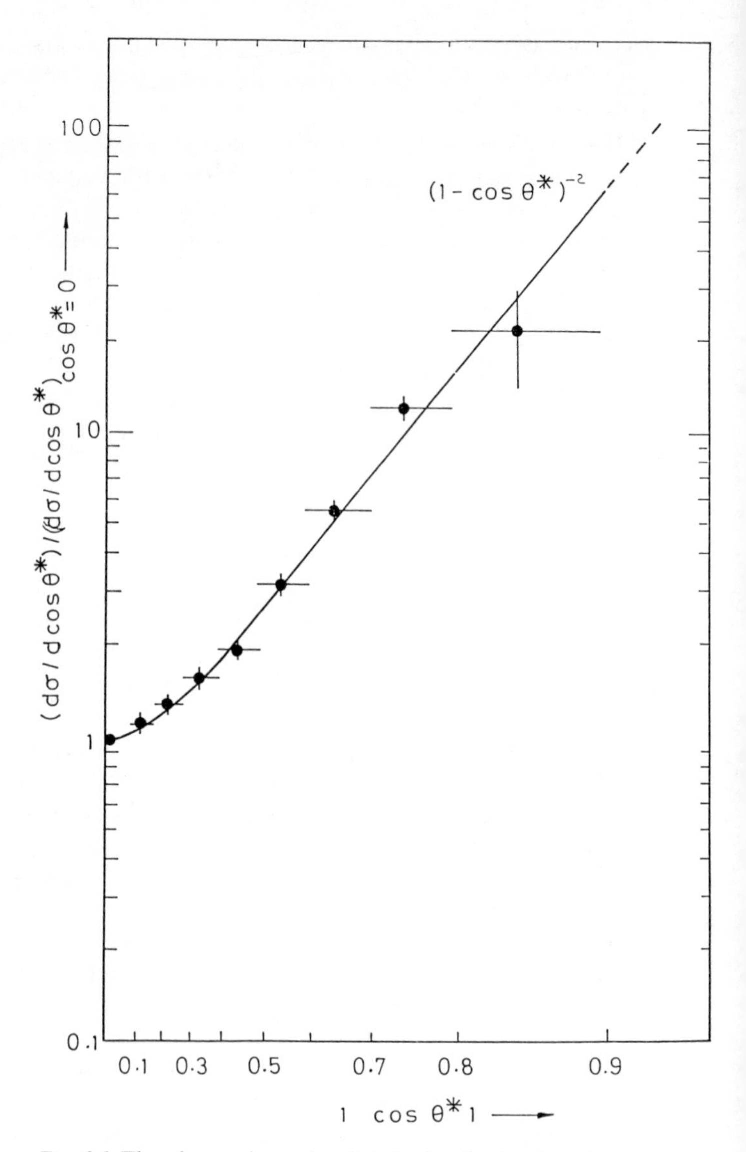

Fig. 6.6 The observed angular distribution for hard parton scattering and the prediction for single-gluon exchange (Arnison *et al.*, *Phys. Lett*. **134B**, 294 (1984)).

Such observations have been made in experiments using e^+e^- collisions (Brandelik (1979), Barber (1979) and Berger (1979)) and latterly $\bar{p}p$ collisions. In e^+e^- collisions the origin of three jets requires a quark or antiquark to emit a gluon

$$e^+e^- \rightarrow q + \bar{q} \quad .$$
$$\hookrightarrow \bar{q} + g$$

The relative rate of three-jet to two-jet production is equal to α_s to a first approximation. Measurements at the colliders (Arnison (1985) and Bartel (1982)) give

$$\bar{p}p: \quad \alpha_s(q^2 = -4000) = 0{\cdot}16 \pm 0{\cdot}02 \pm 0{\cdot}03$$
$$e^+e^-: \alpha_s(q^2 = -900) \ = 0{\cdot}20 \pm 0{\cdot}015 \pm 0{\cdot}03,$$

where the first error term is the statistical contribution and the second is the systematic contribution.

Although the formal explanation of quark behaviour in collisions at large four-momentum transfer is on a sound theoretical footing it has not proved correspondingly possible to explain the binding of quarks and the nucleon properties rigorously from first principles. (Lattice gauge calculations now in progress (1986) show some promise in this direction.) We now concern ourselves with this difficult area. Referring back to the expression for $\alpha_s(q^2)$, we may rewrite it as

$$\alpha_s(q^2) = [B \ln (|q^2|/\Lambda^2)]^{-1}$$

where

$$\Lambda^2 = \mu^2 \exp[-1/B\alpha_s(\mu^2)].$$

$\alpha_s(q^2)$ itself is independent of the reference momentum (μ^2) so that Λ^2 ought to be independent of the reference momentum as well. This may be checked by taking two values of μ and equating the resulting Λ values. After inserting the measured values of $\alpha_s(q^2)$, we obtain

$$\Lambda \simeq 0{\cdot}2 \text{ GeV}.$$

The value of Λ establishes the *scale* of the strong interactions.

For example, the magnitude of Λ is correlated with both the size of nucleons and the range of the strong interactions, $\Lambda^{-1} \sim 10^{-15}$ m.

Quarks are massive particles and the masses of quarks of different flavour differ considerably. Taking the values given in Chapter 2 we see that

$$m_u, m_d \ll \Lambda$$
$$m_s \sim \Lambda$$
$$m_c, m_b, m_t \gg \Lambda.$$

This pattern has the consequence that u- and d-quark masses can be regarded as negligible in QCD calculations, while the c-, b- and t-quark masses do affect QCD calculations. Here, then, is the reason for the identity of proton (uud) and neutron (udd) strong interactions. The strong isospin symmetry (SU(2)), discussed in Chapters 2 and 5 owes its origin to the smallness of the u- and d-quark masses relative to Λ, *not* to their equality. Gell-Mann (1964) and Zweig (1964) extended this idea to propose an SU(3) *flavour* symmetry. This SU(3) symmetry has the three flavours u, d and s as the components of its fundamental representation. It is not such a good symmetry as SU(2) of flavour (strong SU(2)) because the s-quark has a mass comparable to Λ. Finally, there is a vestigial SU(4) symmetry obtained by including u-, d-, s- and c-quarks. The reader should be careful to distinguish between the SU(3) symmetry of colour which underlies the strong force and the *accidental* SU(n) flavour symmetries which exist whenever the quark masses are small or comparable to Λ.

In the short-distance regime where the coupling strength is weak the potential between quark and antiquark is due to one gluon exchange and hence is Coulombic in form ($1/r$). At longer ranges multi-gluon exchange is more likely, leading eventually to confinement of quarks in hadrons. The following argument suggests that this leads to a long-range component of the potential which is linear in r. The colour field between quark and antiquark can be pictured as lines of force which are self-attracting and therefore the lowest energy configuration of this field is a tube of flux of fixed cross-section between quark and antiquark. As these partons move apart the tube gets longer only and hence the potential energy increases linearly with separation. This behaviour contrasts with the behaviour of the lines of an electromagnetic field which fill the whole of the space around two charges. An approximate form of the central potential between a quark and antiquark which shows the correct asymptotic behaviour at short range and is linear at long range is:

$$V(r) = -4\alpha_s/3r + \lambda r$$

with α_s around 0·2 and λ around 0·16. It is also helpful when picturing fragmentation to consider the gluon flux tubes which connect partons at longer ranges. For example, the result of a hard scatter may be that one quark is moving rapidly away from its partner antiquark in a meson. Their separation increases and so does the energy in the flux tube joining them. At some separation it becomes possible to form a quark-antiquark pair in the field. The tube then breaks leaving the parent quark (antiquark) connected by one half of the flux tube to the daughter antiquark (quark). There are now two mesons which can move off independently. Further breakages with multiple meson production are also possible. The separation above which a reduction in energy will result from forming a pair and breaking the flux tube must be about 10^{-15} m.

Chapter 7
The V − A theory of weak interactions

The weak interaction at normal terrestrial energies is indeed very feeble. For example, a μ-lepton has a lifetime of $2{\cdot}2 \times 10^{-6}$ s, a value huge on the nuclear time scale, while neutrinos of energy 4 GeV passing through the earth along a diameter only have a 0·01 per cent chance of interacting on this journey. The weak interaction is also of short range due to the great mass of the field bosons exchanged. Fermi, aware only of the short range, proposed a point contact interaction. This Fermi interaction is ultimately flawed because it predicts that cross-sections rise quadratically with centre-of-mass energy and gives a probability of greater than unity for interactions at high enough energy! The early studies of nuclear β-decay revealed the space-time symmetry properties of the weak interaction ($V-A$). In this context the observation of parity non-conservation was crucial. Studies using particles showed that leptons and quarks feel the same weak interaction. The eigenstates of the quarks as seen by the weak interaction were, however, found to be different from those selected by the strong force (weak and strong isospin). Fermi's model was extended naturally to contain these results and is valuable because apart from the exchange mechanism all the features of the weak aspect of electroweak theory are in place. Electroweak unification is discussed in the next chapter.

7.1 Fermi's model

Fermi (1934) described the weak interaction by analogy with the electromagnetic interaction. For a weak process

$1+2 \rightarrow 3+4$ he wrote the Feynman amplitude (to use modern parlance) for a point-contact interaction:

$$T = \frac{G}{\sqrt{2}} [\bar{u}(4)\gamma_\mu u(2)] [\bar{u}(3)\gamma_\mu u(1)]. \qquad [7.1]$$

Comparison with the single-photon exchange amplitude given in Appendix E and Chapter 4 shows that Fermi's amplitude contains two so-called currents (the terms in brackets) but lacks the propagator which would correspond to an exchange between the currents. We now know that in fact an exchange does occur. However, unlike the photon, the W-boson exchanged in weak interactions has a very large mass (83 GeV/c^2) and the uncertainty principle then ensures that the force has a correspondingly short range. The full amplitude, including the exchange propagator (see section 4.2), is

$$T = (g^2/8) [\bar{u}(4)\gamma_\mu u(2)] \frac{1}{q^2 - M_W^2} [\bar{u}(3)\gamma_\mu u(1)],$$

where g is the strength of the coupling at the vertices and q the four-momentum exchanged. At low enough energies such that $|q^2| \ll M_W^2$ the amplitude approximates to the contact interaction of Fermi's model. Fig. 7.1 contains Feynman diagrams for neutrino scattering and neutron decay in which the W-boson exchange is drawn in explicitly.

The energy release in β-decay is very small compared to the nucleon mass ($\times c^2$), so that the decay nucleon has a low velocity. In this case the nuclear current (Fermi form)

$$\bar{u}(p)\gamma_\mu u(n) = 2M[\phi_p^* 0 0]\gamma_\mu \begin{bmatrix} \phi_n \\ 0 \\ 0 \end{bmatrix}.$$

The form of γ_μ is given explicitly in terms of the Pauli spin matrices in Appendix D. It is straightforward to show that the current is zero for the spatial components ($\mu = 1, 2, 3$) and also zero for the time component ($\mu = 0$) if the neutron and proton have opposite spin components: for example,

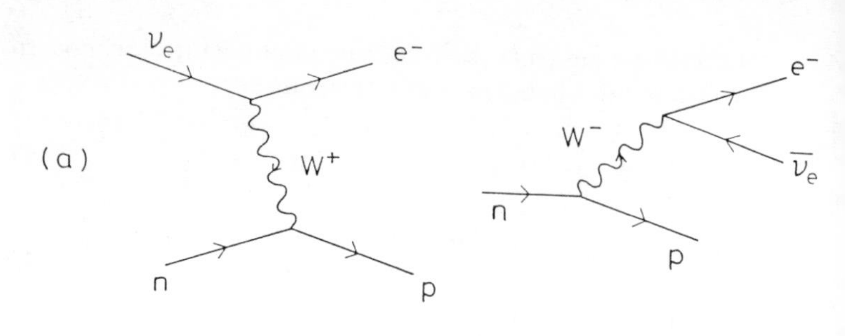

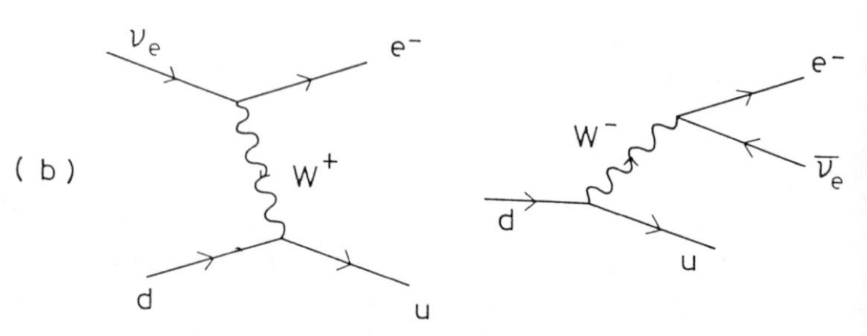

Fig. 7.1 (a) Feynman diagrams for neutrino scattering and neutron decay showing the intermediate vector boson exchange. (b) Feynman diagrams for the neutrino scattering and neutron decay at the quark level.

$$\phi_p^* = [1 \quad 0] \qquad \text{and} \qquad \phi_n = \begin{bmatrix} 0 \\ 1 \end{bmatrix}.$$

Only if the nucleon spin does not flip will there be a non-zero current: for example,

$$\phi_p^* = [1 \quad 0] \qquad \text{and} \qquad \phi_n = \begin{bmatrix} 1 \\ 0 \end{bmatrix}.$$

The fact that the interaction is a point contact means that there is no relative orbital angular momentum. Consequently $\Delta S = \Delta J = 0$ for these allowed decays. Experimentally $\Delta J = 1$

transitions are also observed at rates comparable to the expected $\Delta J=0$ decays and a different current in which γ_μ is replaced by $\gamma_\mu\gamma_5$ seems to be responsible. In the same approximation this second current is non-zero only when there is a nucleon spin flip and $\Delta S=\Delta J=1$. Other forms for the currents also give Lorentz-invariant amplitudes, each with a different choice for the factors replacing γ_μ:

1 scalar (S); γ_μ vector (V);
$(\gamma_\mu\gamma_\nu-\gamma_\nu\gamma_\mu)$ tensor (T); $\gamma_\mu\gamma_5$ axial (A);
γ_5 pseudoscalar (P),

where the form of γ_5 is given in Appendix D.

What permits us to distinguish between these currents experimentally are the expected differences in correlation between the electron and neutrino directions. The decay angular distribution is proportional to $(1+\alpha\beta_e \cos\theta)$, where $\beta_e c$ is the electron velocity and θ is the angle between the electron and neutrino momentum vectors. α is a constant which takes the values -1, $+1$, $+1/3$ and $-1/3$ respectively for S, V, T and A. Any pseudoscalar contribution would be heavily suppressed by kinematic factors. The observed full-strength (allowed) decays all have α equal to either $+1$ or $-1/3$, which limits the coupling to be either vector or axial. Experiments which measure nuclear β-decay rates determine the vector and axial coupling strengths to be very similar, so we replace equation [7.1] by

$$T=\frac{G_V}{\sqrt{2}}[\bar{u}(4)\gamma_\mu u(2)][\bar{u}(3)\gamma_\mu u(1)]$$

$$+\frac{G_A}{\sqrt{2}}[\bar{u}(4)\gamma_\mu\gamma_5 u(2)][\bar{u}(3)\gamma_\mu\gamma_5 u(1)], \qquad [7.2]$$

where $G_V=(1{\cdot}135\pm0{\cdot}003)\times10^{-5}\ \text{GeV}^{-2}$ and $G_A=1{\cdot}26G_V$.

7.2 The $V-A$ theory

Neither the axial nor vector coupling alone can be responsible for the parity-violating effects seen in ^{60}Co decay (Fig. 5.2).

Under the parity transformation ($\mathbf{r} \to -\mathbf{r}$) we have $\gamma_\mu \to -\gamma_\mu$, so that the vector current changes sign, while $\gamma_\mu\gamma_5 \to \gamma_\mu\gamma_5$, so that the axial current is unchanged. In either case the amplitude, being a product of two currents, is unaffected by the parity transformation. It is therefore necessary for the current to include both an axial and a vector piece in order that there should be an interference term in the amplitudes which does change sign under the parity transformation. Then, amplitudes and rates should be affected by the parity transformation. The choices of $V \pm A$ for the current, i.e.

$$\tfrac{1}{2}\bar{u}\gamma_\mu(1 \pm \gamma_5)u,$$

have equal V and A contributions, and these choices would give maximal interference in the amplitude and hence maximal parity violation. At this point the properties of the neutrino become relevant. It is shown in section 4.1 and Appendix D that the masslessness of the neutrino determines that the neutrino has definite handedness: $\frac{1}{2}(1-\gamma_5)$ is the appropriate projection operator for a left-handed neutrino. Evidently the $V-A$ coupling, because it contains this factor, would guarantee that only left-handed neutrinos could couple to other particles. What is the effect of a $V-A$ interaction for other leptons? This is harder to calculate but the result is that the expectation value (mean) of the helicity is proportional to the velocity:

$$\langle \lambda \text{ (lepton)} \rangle = -\beta$$
$$\langle \lambda \text{ (antilepton)} \rangle = +\beta.$$

A similar analysis for the $V+A$ choice shows that these helicity predictions are reversed. Measurements of the polarization of electrons and positrons emitted in nuclear decays have established that $V-A$ is nature's choice. For neutrinos the crucial experiment was performed by Goldhaber, Grodzins and Sunyar (1958) on the K-capture reaction

$$e^- + {}^{150}\text{Eu}(J=0) \to {}^{152}\text{Sm}^*(J=1) + \nu$$
$$\hookrightarrow {}^{152}\text{Sm}(J=0) + \gamma$$

using the apparatus shown in Fig. 7.2. Conservation of angular momentum in the K-capture requires that the Sm* recoil and

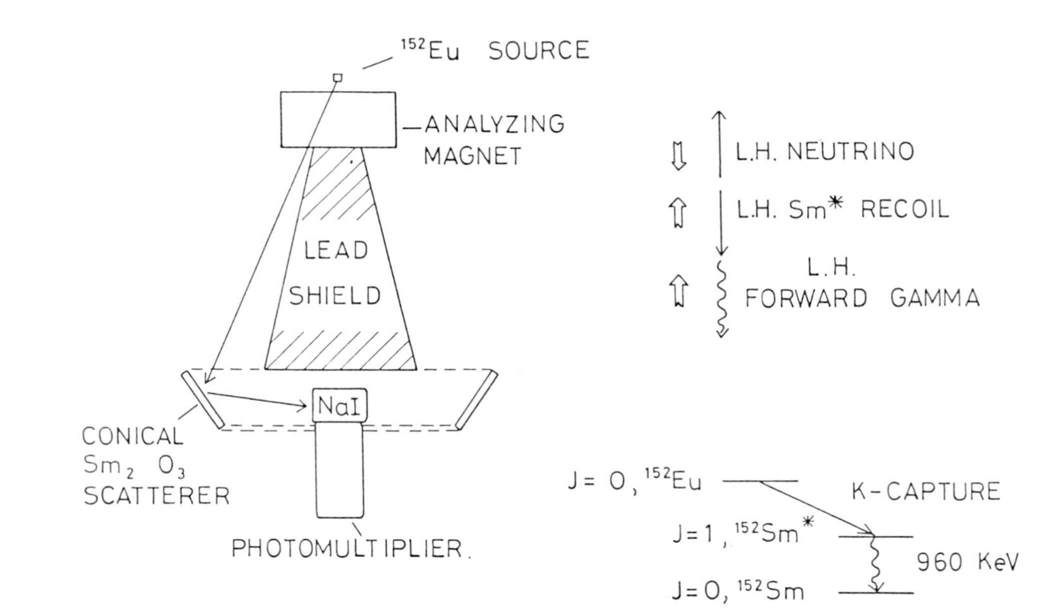

Fig. 7.2 The apparatus used by Goldhaber, Grodzins and Sunyar to measure the helicity of the neutrino emitted following K-capture in ^{150}Eu.

the neutrino have their spins pointing in opposite directions. They therefore have the same sign of helicity: both positive or both negative. The photon carries off the Sm* spin and if the photon is emitted forward the photon will carry the same sign helicity as the recoil and hence as the neutrino. Photon detection used resonance scattering from Sm_2O_3. To establish their polarization the photons were passed through an iron plate magnetized parallel to the photon direction of travel. Electrons in the iron which have their spins opposite to the photon spin direction can easily absorb photons by a spin-flip, but not if their spins are parallel to the photon spin direction. Measurements of the photon transmission for opposite senses of magnetization established the photon polarization to be negative and hence that

$$\lambda\,(\text{neutrino}) = -1.$$

This requires the $V-A$ form of the weak current. The final form for the weak interaction amplitude for nuclear β-decay (which replaces equation [7.2]) is:

$$T = (G_V/\sqrt{2})\,[\bar{u}(4)\gamma_\mu(1-\gamma_5)u(2)]$$
$$[\bar{u}(3)\gamma_\mu(1-\gamma_5 G_A/G_V)u(1)]. \qquad [7.3]$$

Here the first bracket is the pure $V-A$ *lepton* current and the second is the *hadron* current, which is close in form to $V-A$. The question of why the measured values of G_V and G_A are not exactly equal will be discussed later.

The production of μ-leptons from π-meson decay, and their own decays, are both weak processes:

$$\pi^- \to \mu^- + \bar{\nu}_\mu \qquad \tau = 2{\cdot}6 \times 10^{-8}\ \text{s},$$

$$\mu^- \to e^- + \bar{\nu}_e + \nu_\mu \quad \tau = 2{\cdot}2 \times 10^{-6}\ \text{s}.$$

In the first process the parent has spin zero, which means that the spin components of the μ^- and the $\bar{\nu}_\mu$ must cancel. Hence the μ^- must have positive helicity, like the $\bar{\nu}_\mu$. This alignment in the decay makes it easy to obtain polarized μ^--leptons from the decay of π-meson beams. The electrons from μ-lepton decay have been studied very fully and the angular distribution, energy distribution and polarization observed all fit the $V-A$ hypothesis with an amplitude of the form

$$T=(G_{\mu}/\sqrt{2})[\bar{u}(4)\gamma_{\mu}(1-\gamma_5)u(2)][\bar{u}(3)\gamma_{\mu}(1-\gamma_5)u(1)]$$

containing two pure $V-A$ leptonic currents. In particular, the electron helicity is measured to be very close to -1. A calculation of the decay rate gives

$$\Gamma=\frac{G_{\mu}^2 m_{\mu}^5}{192\pi^3}[1-8m_e^2/m_{\mu}^2]. \qquad [7.4]$$

One factor of m_{μ}^3 represents the contribution from phase space with a further m_{μ}^2 coming from the amplitude squared. When the measured decay rate is inserted into this equation it yields $G_{\mu}=(1{\cdot}166365\pm0{\cdot}000016)\times10^{-5}\ \text{GeV}^{-2}$, which is only slightly but nonetheless significantly different from the value of the nuclear G_V. The angular distribution of μ-lepton decay illustrates very well how C and P violation go hand-in-hand:

$$\mu^-: \quad \frac{d\Gamma(\cos\theta)}{\Gamma}=\tfrac{1}{2}(1-\tfrac{1}{3}\cos\theta)\,d(\cos\theta)$$

$$\mu^+: \quad \frac{d\Gamma(\cos\theta)}{\Gamma}=\tfrac{1}{2}(1+\tfrac{1}{3}\cos\theta)\,d(\cos\theta).$$

If we attempt to apply the parity transformation or charge conjugation separately to these distributions we get nonsense. Starting from μ^--decay distribution we would obtain angular distributions of $(1+1/3\cos\theta)$ for μ^--decay and $(1-1/3\cos\theta)$ for μ^+-decay, which conflict with experiment. However, the application of CP leads to a $(1+1/3\cos\theta)$ distribution for μ^+-lepton decays. Evidently CP is a good symmetry although C and P are maximally violated. CP can be seen to generate a right-handed antineutrino from the left-handed neutrino.

The decays of the τ-lepton resemble those of the μ-lepton

$$\tau^- \to \mu^- + \bar{\nu}_{\mu} + \nu_{\tau}$$
$$\tau^- \to e^- + \bar{\nu}_e + \nu_{\tau}.$$

Less data is available than for μ-leptons because of the greater difficulty in producing τ-leptons and their shorter lifetime ($3{\cdot}3 \pm 0{\cdot}4 \times 10^{-13}$ s). Equation [7.4] with the appropriate mass replacements for τ-decay can be used to extract the coupling strength G_τ from the measured τ lifetime. G_τ is found to be consistent with G_μ and the momentum spectra of the decay leptons from τ-decay is also consistent with that expected for a $V-A$ interaction.

The lepton weak-coupling strengths and the nuclear vector-coupling strength are nearly equal, so why is the nuclear axial coupling larger? To restore a single weak-interaction strength via W-exchange requires examination of the nuclear decays in terms of quarks. Fig. 7.1(b) shows the lepton-quark processes underlying the nuclear processes shown in Fig. 7.1(a). At this level the axial and vector coupling constants are equal. When strong-interaction effects are considered the quark is no longer seen to be a single point particle but is continually interacting with other components of the nucleon. Then the currents, electromagnetic or weak, may in principle be affected. However, conservation of charge guarantees that the net charge is the same, whatever emissions take place, so the electromagnetic current

$$\langle \text{nucleon}' | \gamma_\mu | \text{nucleon} \rangle$$

is conserved. This current has the same form as the *vector* component of the weak current, which must therefore be conserved as well. There is no renormalization of the vector coupling constants due to the strong interactions.

Concerning the axial current the situation is less simple. The divergence (rate of change) of the axial current has the same spin-parity (0^-) as the π-meson. Therefore emission of virtual π-mesons by the quark will affect its axial coupling strength. The vector current is said to be conserved, while the axial current is only partially conserved. This asymmetry touches a fundamental problem for QCD: QCD is neither left- nor right-handed, so why are there light 0^- mesons but not a mirror set of 0^+ mesons? Successful calculations of the renormalization to be expected to the axial current have been carried through starting from the phenomenological bag model (Chodos (1974)).

The angular distribution for the simple process $\nu_\mu + e^- \rightarrow \mu^- + \nu_e$ is calculated in Appendix E using the $V-A$ amplitude. Taking the result we have in the CM frame

$$d\sigma/d\Omega^* = G^2 s/4\pi^2,$$

which is isotropic. The total cross-section is thus

$$\sigma = G^2 s/\pi$$

and would rise indefinitely as the energy increases. This illustrates the fundamental problem of the Fermi model: eventually cross-sections become so large that the probability of an interaction exceeds unity! It requires inclusion in the amplitude of the massive-boson exchange between the currents to restrain this rise in cross-section.

The expression for cross-section given above has the form

$$\sigma \sim G^2 s$$

while equation [7.4] for a decay rate is of the form

$$\Gamma \sim G^2 M^5.$$

These two general forms are easy to obtain using dimensional arguments plus the fact that the rates of weak processes are proportional to the Fermi coupling strength squared. In the absence of complications from spin dependence these two expressions are quite useful for calculating the relative rates of similar processes.

7.3 Neutral K-mesons

K-mesons have the same spin-parity, 0^-, as the π-mesons but are heavier because one of the component light u- or d-quarks is replaced by an s-quark. The K^0-meson ($\bar{s}d$) has mass 497·7 MeV compared to 135·0 MeV for the π^0-meson ($(d\bar{d} - u\bar{u})/\sqrt{2}$). The K-mesons decay weakly due to the underlying weak process $s \rightarrow u + W^-$ analogous to $d \rightarrow u + W^-$ in neutron decay (see Fig. 7.1). Sometimes the decay results in lepton emission and sometimes not. A typical quark level diagram contributing to non-leptonic decay is drawn in Fig. 7.3. The

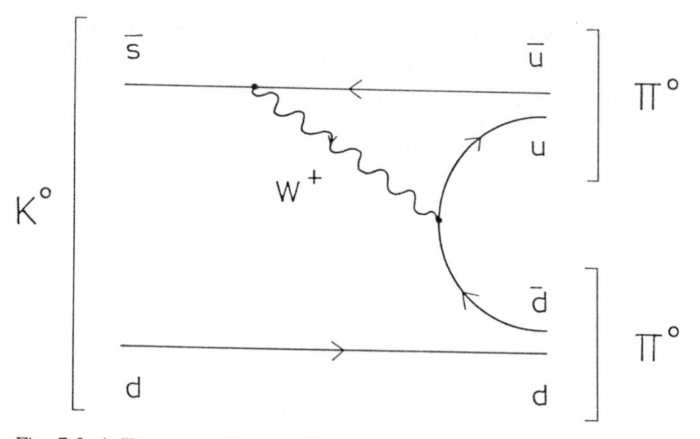

Fig. 7.3 A Feynman diagram for a weak process underlying the decay $K^0 \rightarrow \pi^0 + \pi^0$.

$K^+(u\bar{s})$ and $K^0(d\bar{s})$ form a strong isospin doublet. Their antiparticles are $K^-(\bar{u}s)$ and $\bar{K}^0(\bar{d}s)$ respectively. The neutral K^0-mesons have the unexpected property of having two different lifetimes which depend on the decay mode:

$$K_S^0 \rightarrow \pi^+ + \pi^- \qquad \tau_S = 0{\cdot}89 \times 10^{-10}\ \text{s} \qquad \text{(Short-lived)}$$
$$K_L^0 \rightarrow \pi^+ + \pi^0 + \pi^- \qquad \tau_L = 5{\cdot}183 \times 10^{-8}\ \text{s.} \qquad \text{(Long-lived)}$$

Quantum mechanics requires that the two modes must be orthogonal eigenstates because they have different lifetimes. What otherwise distinguishes between the two modes is their behaviour under the product operation *CP*. Both final states have angular momentum zero, so that their parities are just the products of the π-meson parities: $P(K_S^0)$ is $+1$ and $P(K_L^0)$ is -1. In this instance the *C* operation is equivalent to interchanging the charged π-mesons. Because the orbital angular momentum is zero this interchange leads to no change in the wavefunction. Thus $CP(K_S^0)$ is $+1$ and $CP(K_L^0)$ is -1. It was seen earlier that *CP* is conserved in weak interactions so that these decay modes are orthogonal eigenstates and may have different lifetimes. The very much larger kinetic energy available in the decay to two π-mesons means that this mode has the more rapid decay rate.

The strong-interaction eigenstates are K^0 and $\bar{K}^0$ and these have opposite flavour (strangeness +1 and −1 respectively). In terms of these states the weak-interaction eigenstates are

$$K_S^0 = (K^0 + \bar{K}^0)/\sqrt{2}, \quad K_L^0 = (K^0 - \bar{K}^0)/\sqrt{2}.$$

Conversely

$$K^0 = (K_S^0 + K_L^0)/\sqrt{2}, \quad \bar{K}^0 = (K_S^0 - K_L^0)/\sqrt{2}.$$

This distinction between strong and weak eigenstates leads to the curious effect of strangeness oscillation in a beam of initially pure K^0 mesons produced in a reaction such as

$$\pi^- + p \to K^0 + \Lambda^0.$$

Within a few times 10^{-10} s the K_S^0 component has decayed leaving only the K_L^0 component. This is a superposition of K^0 and $\bar{K}^0$, which means that the beam has changed its flavour! It is partly of positive strangeness and partly of negative strangeness. In order to demonstrate this behaviour the frequency of the decays

$$K^0 \to \pi^- + e^+ + \nu_e$$
$$\bar{K}^0 \to \pi^+ + e^- + \bar{\nu}_e$$

(which label the strangeness of the parent by the lepton sign) is monitored with position along the beam. We will analyse the development in time of an initially pure K^0 beam in detail because its measurement yields the most precise information we possess concerning the mass differences between a particle and its antiparticle. The initial beam at time zero is

$$|0\rangle = 1/\sqrt{2}[|K_S^0\rangle + |K_L^0\rangle]$$

which after a time t becomes (Appendix B)

$$|t\rangle = 1/\sqrt{2}[|K_S^0\rangle \exp(-im_S t - \Gamma_S t/2) + |K_L^0\rangle \exp(-im_L t - \Gamma_L t/2)],$$

where m is the mass and Γ the decay rate.

Notice that the mean lifetime $\tau = 1/\Gamma$. At times short compared to τ_L this reduces to:

$$|t\rangle = \exp(-im_S t)/\sqrt{2}[|K_S^0\rangle \exp(-\Gamma_S t/2) + |K_L^0\rangle \exp(-i\Delta m t)],$$

where $\Delta m = m_L - m_S$. The amplitude for K^0 is given by

$$A(t) = \langle K^0|t\rangle = \exp(-im_S t)/2[\exp(-\Gamma_S t/2) + \exp(-i\Delta m t)].$$

Thus the intensity of the K^0-meson component at time t is

$$K^0(t) = |A(t)|^2 = (1/4)[\exp(-\Gamma_S t) + 2\exp(-\Gamma_S t/2)\cos(\Delta m t) + 1].$$

This starts at unity and as time passes shows an oscillatory component which depends on the mass difference. The $\bar{K}^0$-meson intensity is

$$\bar{K}^0(t) = (1/4)[\exp(-\Gamma_S t) - 2\exp(-\Gamma_S t/2)\cos(\Delta m t) + 1],$$

which initially rises from zero and then shows oscillations of the same frequency. After times much longer than τ_S but still short compared to τ_L, the K_L^0 component alone remains and $K^0(t) = \bar{K}^0(t) = 1/4$.

Gjesdal *et al.* (1974) find, from fitting the observed oscillations, that:

$$m_L - m_S = (3{\cdot}521 \pm 0{\cdot}014) \times 10^{-12}\ \text{MeV}.$$

This is remarkably close to the value of Γ_S, *viz.* $7{\cdot}376 \times 10^{-12}$ MeV. The similarity arises because both quantities are determined by the coupling

$$K_S^0 \rightleftharpoons \pi + \pi.$$

Following this line of reasoning it can be argued that the only mechanism available to generate a difference in mass between K^0 and $\bar{K}^0$ would be the second-order process:

$$K^0 \rightleftharpoons \pi + \pi \rightleftharpoons \bar{K}^0.$$

Any mass difference which might be generated would be much

smaller than $(m_L - m_S)$; detailed consideration puts the limit at

$$|m(K^0) - m(\bar{K}^0)| < 3 \times 10^{-16} \text{ MeV}.$$

This result provides the most stringent test on the exact equality of particle and antiparticle masses required by *CPT* invariance.

The study of neutral K-mesons produced a further surprise. In 1964 Christiansen, Cronin, Fitch and Turlay discovered decays to two π-mesons in a K_L^0 beam, at a low rate, now measured to be (0·094 ± 0·018) per cent. This decay is necessarily a *CP* violating process, and if *CPT* is to remain a valid symmetry then this in turn implies violation of time-reversal invariance. Despite sustained searches, no other evidence for *CP* violation has been seen outside the neutral K-meson system. In particular, any violation of *P* or *T* could lead to some contribution to a neutron electric dipole moment. Current measurements impose an upper limit of 10^{-23} e cm on this quantity.

7.4 The Cabibbo hypothesis and its extension

The rate of hadron weak decay depends strongly on the change in flavour of the quark involved. An excellent example is provided by the alternative decays of the Σ^--hyperon (with the quark content listed below):

$$\Sigma^- \to n + e^- + \bar{\nu}_e$$
$$(dd)s \to (dd)u + e^- + \bar{\nu}_e$$
$$\Sigma^- \to \Lambda^0 + e^- + \bar{\nu}_e$$
$$(ds)d \to (ds)u + e^- + \bar{\nu}_e.$$

If the decays had the same coupling strength then the decay rates would be in the ratio given by fifth powers of the mass difference in each decay (see equation [7.4]), i.e.

$$\Gamma(\Sigma^- \to n)/\Gamma(\Sigma^- \to \Lambda^0) = (M_\Sigma - M_n)^5/(M_\Sigma - M_\Lambda)^5$$
$$= 300,$$

whereas the observed ratio is 18 ± 2. Comparing the vector coupling strengths:

$$\text{muon decay} \quad G_V(\mu \rightarrow e) = 1{\cdot}166365\ (\pm 16) \times 10^{-5}\ \text{GeV}^{-2}$$

$$\text{nuclear decay} \quad G_V(d \rightarrow u) = 1{\cdot}135\ (\pm 3) \times 10^{-5}\ \text{GeV}^{-2}$$

$$\text{strangeness changing} \quad G_V(s \rightarrow u) = 0{\cdot}25\ (\pm 2) \times 10^{-5}\ \text{GeV}^{-2}.$$

This comparison seems at first sight to rule out a universal coupling of quarks and leptons to the weak vector boson. However, Cabibbo saw that it was possible to preserve this desirable feature. He hypothesized that the quark which couples to the u-quark via the weak interaction is not a pure d-quark or s-quark but the linear superposition:

$$d' = d \cos \theta_C + s \sin \theta_C$$

with θ_C being a mixing angle now called the Cabibbo angle. It then follows that

$$G_V(d \rightarrow u) = G_V(\mu \rightarrow e) \cos \theta_C = 0{\cdot}975\ G_V(\mu \rightarrow e)$$
$$G_V(s \rightarrow u) = G_V(\mu \rightarrow e) \sin \theta_C = 0{\cdot}220\ G_V(\mu \rightarrow e),$$

where we have inserted the measured value of θ_C (0·222). Cabibbo's hypothesis accounts for both the lower rate of strangeness-changing decays and for the puzzling small discrepancy between $G_V(\mu \rightarrow e)$ and $G_V(d \rightarrow u)$. The same results carry over to the axial coupling strength with an identical value of θ_C.

The combination orthogonal to the d′-quark ($s \cos \theta_C - d \sin \theta_C$) is called s′. This couples through the weak interaction to the c-quark, which has mass around 1500 MeV. Hadrons containing the c-quark are called charmed particles. Of these the pseudoscalar D^+(1869) and D^0(1865) mesons have lifetimes of $0{\cdot}92 \pm 0{\cdot}12$ and $0{\cdot}44 \pm 0{\cdot}07$ ps, which indicates that they decay weakly. Decays to strange particles and to leptons are common, 60 per cent and 38 per cent respectively for D^+(1869) which has quark content $c\bar{u}$. This demonstrates the importance of the quark coupling,

$$c \rightarrow s + l^+ + \nu_l.$$

Determinations of D-meson lifetimes rely on measurements of path length between the production and decay vertices in bubble chambers, in emulsions or in electronic detectors. The paths are short: at 19 GeV a $D^+(1869)$ meson will travel only a distance

$$l = \tau\beta\gamma c = 0{\cdot}28 \text{ cm}.$$

Kobayashi and Maskawa pointed out in 1973 that *CP* violation could be accommodated in the standard model if there were three quark generations. The Cabibbo mixing generalizes so that the flavour weak coupling is expressed by the matrix

$$[\mathrm{u} \quad \mathrm{c} \quad \mathrm{t}] \begin{bmatrix} U_{\mathrm{ud}} & U_{\mathrm{us}} & U_{\mathrm{ub}} \\ U_{\mathrm{cd}} & U_{\mathrm{cs}} & U_{\mathrm{cb}} \\ U_{\mathrm{td}} & U_{\mathrm{ts}} & U_{\mathrm{tb}} \end{bmatrix} \begin{bmatrix} \mathrm{d} \\ \mathrm{s} \\ \mathrm{b} \end{bmatrix}.$$

The matrix is unitary (if there are no further flavours to be discovered) and all the components can be described using three Cabibbo-like angles plus a phase which parametrizes the *CP* violation.

The b-quark is massive (around 5 GeV). Hadrons containing b-quarks have been produced in e^+e^- annihilations as we shall discuss in Chapter 10. Leptonic decays account for about 23 per cent of all b-quark decays. The observed energy spectrum of the decay leptons does not extend above 2·5 GeV and this signifies that decays are mainly to c-quarks rather than to the much lighter u-quarks. Preliminary measurements put the lifetime of the b-quark at about 1·5 ps. At present (1986) the existence and decay modes of the t-quark are not firmly established. The limited data available on heavy-quark decays is consistent with the quark weak couplings being $V-A$ and of equal strength with the lepton weak couplings. The Kobayashi-Maskawa matrix provides a useful parametrization of how the mass eigenstates (d, s, b) mix to form the weak eigenstates (d′, s′, b′). A full theoretical explanation of its content and significance is lacking at present because we have no rationale for the existence of several generations of leptons and quarks. It is also uncertain at the time of writing whether all the *CP* violation observed can be accommodated by the single phase in the mass matrix.

All weak processes discussed thus far share a common origin, that is the $(V-A)$ coupling of the charged W-boson to a lepton or quark current. Fig. 7.4 shows the couplings which possess equal strength. In the quark-antiquark case the colours must match because the weak current is 'colourless'. A neat example of the distinction provided by the colour quantum number occurs in the relative rates of the hadronic and leptonic

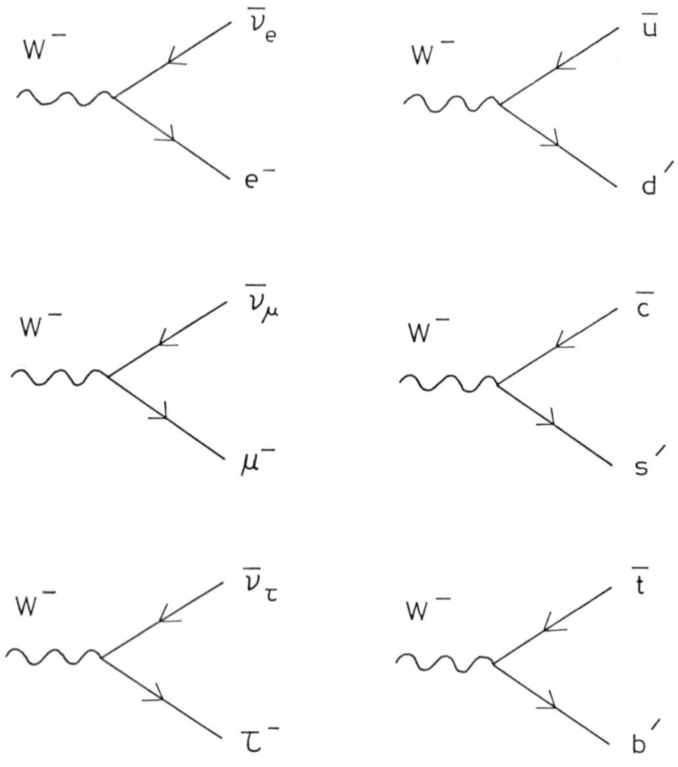

Fig. 7.4 The equivalent strength weak coupling vertices of the standard model.

τ-decays. The energetically allowed decays are:

$$\tau^- \to \nu_\tau + W^- \\ \quad \hookrightarrow \begin{cases} e^- + \bar{\nu}_e \\ \mu^- + \bar{\nu}_\mu \\ \bar{u} + d'. \end{cases}$$

Decays to e^- and μ^- are expected to occur at equal rates while the decays to hadrons are expected to be three times as frequent because each choice for the quark's colour must be counted as a separate quantum state. The respective branching fractions of τ-decay are in fact measured to be 16·5, 18·5 and 65 per cent.

In summary the weak interaction has a universal strength for leptons and quarks. The weak currents appearing in the amplitudes connect either a pair of leptons or a pair of quarks. These currents are all $V-A$ in form and lead naturally to parity and C-parity violation in weak processes. The weak eigenstates of the quarks are linear superpositions of the mass eigenstates.

Chapter 8
Electroweak unification

The Fermi theory of weak interactions leads to an unacceptable growth of cross-sections as beam energy increases. This particular problem is solved by the introduction of a massive vector boson exchange between currents. Amplitudes then contain the factor $g^2/8(q^2 - M_W^2)$, where g is the boson coupling constant, instead of $G/\sqrt{2}$. As the CM energy increases the q^2 factor dominates M_W^2 and this leads to a restraint in cross-section increase. Notwithstanding this improvement the theory is not renormalizable. Glashow (1961), Weinberg (1967) and Salam (1968) proposed a gauge theory which is fully renormalizable and has the added advantage that it describes both weak and electromagnetic forces in a unified manner: this is the celebrated electroweak unification. The theory predicts the existence of a neutral partner (Z^0) to the $W^{\pm}$-bosons. The validity of the electroweak model has been confirmed in a series of remarkable experiments.

8.1 Weak isospin and hypercharge

Before introducing the symmetries which underlie the electroweak model we make one or two general remarks. Each SU(n) internal symmetry has an n-fold fundamental representation, so it is likely that the fundamental particles should form such a representation. Furthermore the force resulting from gauging such a symmetry will not be capable of distinguishing between the n different members of the fundamental representation. With these points in mind we recall that the leptons and quarks all undergo weak interactions and appear in definite pairings in the currents. Weak interactions are of universal strength; G is the same whether the current is leptonic $(\bar{l}\gamma_\mu(1-\gamma_5)l')$ or quarkic $(\bar{q}\gamma_\mu(1-\gamma_5)q')$.

This association of both quarks and leptons into pairs with identical weak couplings makes it plausible that the relevant symmetry underlying the weak force is SU(2): i.e. weak isospin symmetry. Experiment shows that at sufficiently high velocity only left-handed particles or their right-handed antiparticles feel the weak force, which means that the symmetry must be restricted to $SU(2)_L$ rather than SU(2). In this view only the left-handed fermions form doublets:

$$(\mathrm{u}, \mathrm{d}')_L \quad (\mathrm{c}, \mathrm{s}')_L \quad (\mathrm{t}, \mathrm{b}')_L$$
$$(\mathrm{e}^-, \nu_e)_L \quad (\mu^-, \nu_\mu)_L \quad (\tau^-, \nu_\tau)_L,$$

where the use of primes indicates that we are referring to the weak eigenstates of the down, strange and beauty quarks as described in the previous chapter. All the right-handed fermions are singlets:

$$(\mathrm{u}, \mathrm{d}, \mathrm{c}, \mathrm{s}, \mathrm{t}, \mathrm{b}, \mathrm{e}^-, \mu^-, \tau^-)_R.$$

u_L and d'_L have weak isospin (t) of $\frac{1}{2}$ and third components (t_3) of $+\frac{1}{2}$ and $-\frac{1}{2}$ respectively; u_R and d_R both have a weak isospin equal to zero. Written as weak-isospin state vectors

$$u_L = |\tfrac{1}{2}, \tfrac{1}{2}\rangle, \qquad d'_L = |\tfrac{1}{2}, -\tfrac{1}{2}\rangle,$$
$$u_R = |0, 0\rangle \quad \text{and} \quad d_R = |0, 0\rangle.$$

The reader should take notice of the following significant difficulty in using a local gauge theory to explain weak interactions. There would certainly be gauge bosons to carry the weak force but they, like the photon, would be massless. Now this clearly contradicts what was explained earlier: namely that the W-boson should be massive in order that at low energies the interaction tends to the point-contact form. Fortunately a mechanism which supplies mass to the field particles was discovered by Higgs (1964). It requires spontaneous symmetry breaking, i.e. a symmetry of the dynamics (equations of motion) is absent from the lowest energy state.

Ferromagnetism provides one familiar example of symmetry breaking. The spin-spin forces acting between the atoms do not depend on the alignment of the specimen in space; nevertheless the lowest energy state is one in which the atomic spins are aligned in a particular direction. In particle theory the analogous situation is to have a vacuum state which lacks some symmetry of the theory. Symmetry transformations then change one vacuum into another equivalent vacuum. These transformations change neither energy nor angular momentum: quantization of the theory leads to the emission or absorption of massless scalar particles during these transitions. Such particles are called Goldstone bosons.

At first sight the introduction of spontaneous symmetry breaking appears to make a bad situation worse—in addition to the unobserved massless gauge bosons, there are now unobserved massless scalar bosons. Higgs made the happy discovery that for gauge theories, and for gauge theories *alone*, the gauge and scalar bosons mix to give massive vector bosons. The resulting theory then has the necessary property of being renormalizable. Note that massless gauge bosons start with only two helicity components; like photons they can be right- or left-circularly polarized but never have helicity zero. Through the Higgs mechanism each weak gauge boson picks up an helicity zero component *and* acquires mass by absorbing a Goldstone boson. The details of the Higgs mechanism are open to discussion. In what follows here the simplest view is adopted, and questions such as whether the Higgs boson is composite are ignored.

The final ingredient of the electroweak model is to introduce what is called a weak hypercharge symmetry U(1). Weak hypercharge (y) and weak isospin are linearly related to the electromagnetic charge (Q).

$$Q = t_3 + y/2.$$

Left-handed quarks have a weak hypercharge of $+\frac{1}{3}$; left-handed leptons have a weak hypercharge of -1; right-handed fermions have Q equal to $y/2$. It is the compound symmetry $SU(2)_L \otimes U(1)$ which is gauged to produce a unified description of weak and electromagnetic interactions. The W^+, W^0 and W^- are the field bosons which emerge from gauging $SU(2)_L$ and B is the neutral field boson from U(1). There are therefore two coupling strengths: g for a W-boson coupling to fermions and g' for a B-boson coupling to fermions, SU(2) is a non-Abelian symmetry, so it is possible not only for field bosons to couple to fermions (Fig. 8.1(a)) but also for them to couple to one another at three or four boson vertices (Fig. 8.1(b) and (c)). The effect of the Higgs mechanism is for the primitive field bosons to absorb the Goldstone bosons and to mix giving the set of observable field bosons:

$$W^+, W^-$$
$$Z^0 = W^0 \cos\theta_W - B \sin\theta_W$$
$$A = W^0 \sin\theta_W + B \cos\theta_W.$$

Among these, W^+, W^- and Z^0 have acquired mass but the photon (A) remains massless; θ_W is the Weinberg mixing angle. Of course, the coupling strength of the photon is well known and constrains the values of g and θ_W,

$$e = g \sin\theta_W = gg'/(g'^2 + g^2)^{\frac{1}{2}}. \qquad [8.1]$$

Further, the masses of the $W^{\pm}$ and Z^0 are related to the vacuum amplitude of the Goldstone field (η):

$$M(\text{W-boson}) = g\eta/\sqrt{2}$$
$$M(\text{Z-boson}) = M(\text{W-boson})/\cos\theta_W. \qquad [8.2]$$

At low energies the electroweak amplitude involving charged W-boson exchange collapses to the Fermi point-interaction amplitude, so there exists a connection between g and G:

$$G = g^2/(4\sqrt{2}M_W^2). \qquad [8.3]$$

These last four expressions relate g, M_W, M_Z and θ_W to measured quantities but they are not enough to fix their absolute values.

The electroweak unification makes two distinctive predictions. Firstly that there exist *massive* vector bosons which transmit the weak force between fermions. Secondly that the

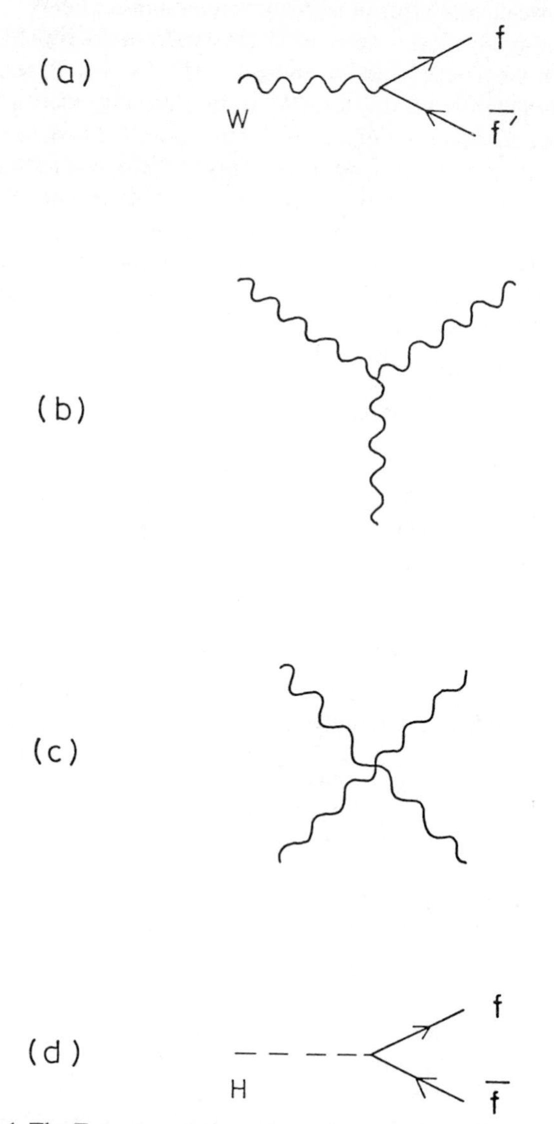

Fig. 8.1 The Feynman vertices of the electroweak model: (a) the fermion gauge-boson coupling; (b), (c) the three and four gauge boson vertices; and (d) the Higgs-fermion coupling.

weak interaction involving the exchange of a *neutral* boson must occur. The cases of the 'charged' and the unexpected 'neutral-current' interactions are compared in Fig. 8.3. There is the charged-current process

$$\nu_\mu + e^- \rightarrow \mu^- + \nu_e$$

via W exchange, but now also, the neutral-current process

$$\nu_\mu + e^- \rightarrow \nu_\mu + e^-$$

via Z^0 exchange.

After this conceptual introduction a more detailed presentation of the electroweak model is made along the lines adopted for QED and QCD. The reader who wishes to avoid this detail can pass straight to section 8.3 and read about the experimental confirmation of electroweak unification.

8.2 Electroweak unification based on $SU(2)_L \otimes U(1)$

The basic requirement of the model is that the Dirac equation be made invariant under the local $SU(2)_L \otimes U(1)$ transformations. These are built from infinitesimal rotations $\alpha_i(x)$ in weak-isospin space and $\beta(x)$ in weak-hypercharge space which depend on the position x in space-time. Then the fermion (quark or lepton) spinor transforms as follows:

$$\psi \rightarrow (1 + ig\alpha_i t_i^L + i(g'/2)\beta y)\psi.$$

Here y is the weak hypercharge and t_i^L the left-handed weak isospin,

$$t_i^L = (\tau_i/2)\,[\tfrac{1}{2}(1-\gamma_5)],$$

where τ_i is a Pauli matrix acting in weak-isospin space. $\alpha_i(x)$ and $\beta(x)$ vary across space-time, so in order to maintain gauge invariance the derivative must be replaced by the covariant derivative

$$D_\mu = \partial_\mu + igt_i^L W_{i\mu}(x) + i(g'/2)yB_\mu(x),$$

where g and g' are the boson-fermion coupling constants for weak isospin and hypercharge. They are analogous to g_s the

QCD coupling constant and e in QED. $W_{i\mu}(x)$ is the $SU(2)_L$ gauge field, being a vector in space-time (index μ) and a vector also in isospin space (index i). The combinations of definite charge are $W^{\pm}=(W_1 \pm W_2)/\sqrt{2}$, $W^0 = W_3$. B_μ is the U(1) gauge field, being a vector in space-time and a weak-isospin scalar. It is thus neutral. The Dirac equation becomes

$$[i\gamma_\mu \partial_\mu - g\gamma_\mu t_i^{\mathrm{L}} W_{i\mu} - (g'/2)y\gamma_\mu B_\mu - m]\psi = 0.$$

The extra terms in the covariant derivative give rise to the 'minimal weak couplings' between the gauge fields and the fermions, exactly as in the cases of QED and QCD. When quantized the theory will have corresponding weak-interaction vertices. These couplings are $g\gamma_\mu t_i^{\mathrm{L}} W_{i\mu}$ and $(g'/2)y\gamma_\mu B_\mu$ for the fermion-boson vertices of the type shown in Fig. 8.1(a). Triple and quadruple boson couplings occur (Fig. 8.1(b) and 8.1(c)) in the same way as for QCD because the symmetry is non-Abelian, but unlike QED.

We now discuss how the components of W_μ and B_μ acquire mass by the Higgs mechanism. The simplest form for the Goldstone field is a weak-isospin doublet:

$$\phi = \begin{bmatrix} \phi_1 \\ \phi_2 \end{bmatrix} \quad \text{and} \quad \phi^+ = [\phi_1^*, \phi_2^*],$$

where $\phi_1(\phi_2)$ is positive (neutral). This has an energy density for which the most general expression containing only fields and their first derivatives is

$$E = D_\mu \phi^+ D_\mu \phi + \mu^2 \phi^+ \phi + \lambda(\phi^+ \phi)^2.$$

The above form can be partially understood using dimensional arguments. Energy has dimension GeV and volume has dimension GeV^{-3}, so that energy density has dimension GeV^4. Derivatives have dimension GeV and boson fields have dimension GeV. For example, the electromagnetic field energy has the (dimensional) form $(\partial A)^2 \times$ volume. Therefore in the above expression μ has dimension GeV (a mass) and λ is dimensionless (a coupling constant). Higher-order terms in ϕ such as $\eta(\phi^* \phi)^3$ are disallowed because the theory would not be renormalizable. Finally the product $\phi^*\phi$ must be present everywhere otherwise the *absolute* phase of ϕ would become significant.

In the state of lowest energy (vacuum) ϕ is everywhere a constant (ϕ_v) and its value is very different for the two possibilities $\mu^2>0$ and $\mu^2<0$. λ being a coupling constant is positive. If $\mu^2>0$ the Goldstone field in the vacuum vanishes, i.e. $\phi_v\equiv 0$; but if $\mu^2<0$, $\phi_v^+\phi_v=-\mu^2/2\lambda\equiv\eta^2$, i.e. there is a non-zero value of ϕ in the vacuum (see Fig. 8.2). In the second case many alternative vacua exist differing only in the phase of ϕ_v. Massless Goldstones are emitted and absorbed in transitions between these vacua. Although the energy equation is

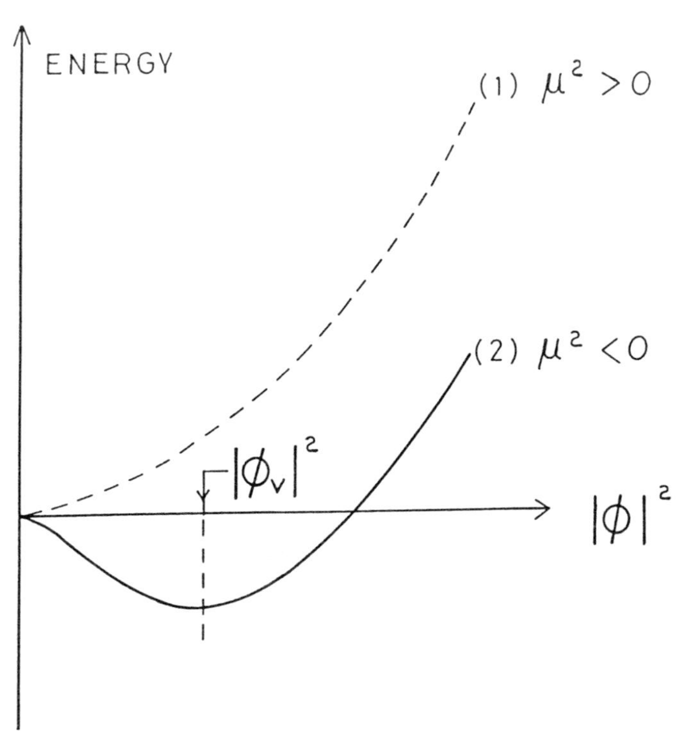

Fig. 8.2 The energy in the scalar field (1) in the case of a vanishing vacuum field and (2) in the case of a non-vanishing vacuum field ϕ_v.

symmetric under weak-isospin rotations the vacuum is not; ϕ_v has some definite isospin. This is just the situation called spontaneous symmetry breaking. We are at liberty to select the local isospin axis and phase and we make the choices in such a way that the states are neutral and the vacuum is also real:

$$\phi_v = \begin{bmatrix} 0 \\ \eta \end{bmatrix} \quad \phi = \begin{bmatrix} 0 \\ \eta + \sigma(x)/\sqrt{2} \end{bmatrix}.$$

η is real and $\sigma(x)/\sqrt{2}$ is the piece over and above the vacuum and is therefore the observed field. Both have $t_3 = -\frac{1}{2}$ and $y = 1$. Then the energy density of the Goldstone field becomes, replacing ϕ by $(\eta + \sigma(x)/\sqrt{2})$ and expanding the covariant derivative,

$$E = \tfrac{1}{2}\partial_\mu \sigma^* \partial_\mu \sigma + \tfrac{1}{4} g^2 \eta^2 [W_1^2 + W_2^2]$$
$$+ \tfrac{1}{4}\eta^2 [g W_3 - g' B]^2 + \text{terms in the Higgs field.}$$

There are two new features to note: terms quadratic in W and B have appeared and not their derivatives. In the quantized theory the field particles therefore carry a mass determined by the coefficients of these quadratic terms; thus M_w^2 is $g^2\eta^2/2$. The terms in the neutral fields W_3 and B are mixed so we need to diagonalize this expression in order to identify the observable fields. The result is

$$Z = W^0 \cos\theta_W - B \sin\theta_W$$
$$A = W^0 \sin\theta_W + B \cos\theta_W$$

with the Weinberg angle θ_W given by $\cos\theta_W = g/(g^2 + g'^2)^{\frac{1}{2}}$ and $\sin\theta_W = g'/(g^2 + g'^2)^{\frac{1}{2}}$. Then the field energy density (ignoring terms in the Higgs field) is

$$E = \tfrac{1}{4} g^2 \eta^2 [(W^+)^2 + (W^-)^2 + Z^2/\cos^2\theta_W].$$

There is no quadratic term in A, which means that this field has *massless* quanta and is to be identified as the electromagnetic field. All the remaining fields have acquired mass:

$$M(\text{W-boson}) = g\eta/\sqrt{2}$$
$$M(\text{Z-boson}) = M(\text{W-boson})/\cos\theta_W.$$

The $W^{\pm}$ and Z^0 bosons have each absorbed one of the four (real plus imaginary) components of the original Goldstone. Its residual component has become massive and is called a Higgs boson: the model gives no prediction for the Higgs boson mass. Higgs bosons are expected to couple to fermions (Fig. 8.1(d)) with a strength that is proportional to the fermion mass. Finally, rewriting the Dirac equation in terms of the observable bosons (but without the Higgs) gives

$$[i\gamma_\mu \partial_\mu - \gamma_\mu \{(g/\sqrt{2})(W_\mu^- t_+^L + W_\mu^+ t_-^L) + g \sin\theta_W Q A_\mu + (g/\cos\theta_W)(t_3^L - Q\sin^2\theta_W) Z_\mu\} - m]\psi = 0,$$

where $t_\pm^L = t_1^L \pm i t_2^L$ and $Q = t_3^L + y/2$. The coupling coefficient of the electromagnetic field A_μ has to be the charge so that

$$e = g \sin\theta_W.$$

This is the crucial equation which links the vector-boson masses and the coupling strengths. Another term in the covariant derivative appears in the Dirac equation as the weak charged-current interaction with coupling

$$\gamma_\mu (g/\sqrt{2}) t_\pm^L = (g/2\sqrt{2}) t_\pm \gamma_\mu (1 - \gamma_5),$$

where t_+ would be operative for $\nu \rightarrow l$ and t_- would be operative for $l \rightarrow \nu$. What is completely new is the weak neutral-current interaction involving the massive Z^0 boson. This interaction will lead to unexpected processes involving neutrinos, such as

$$\nu_\mu + e^- \rightarrow \nu_\mu + e^-.$$

Fig. 8.3 compares charged- and neutral-current interactions.

In order to calculate the Fermi coupling constant we consider the charged-current reaction:

$$\nu_\mu + e^- \rightarrow \mu^- + \nu_e.$$

The current for $\nu_\mu \rightarrow \mu^-$ is

$$\langle \tfrac{1}{2}, +\tfrac{1}{2} | \bar{u}(\mu) \gamma_\mu [g/\sqrt{2} t_+^L] u(\nu) | \tfrac{1}{2}, -\tfrac{1}{2} \rangle,$$

where the weak-isospin components are displayed explicitly. The effect of the isospin stepping operator is

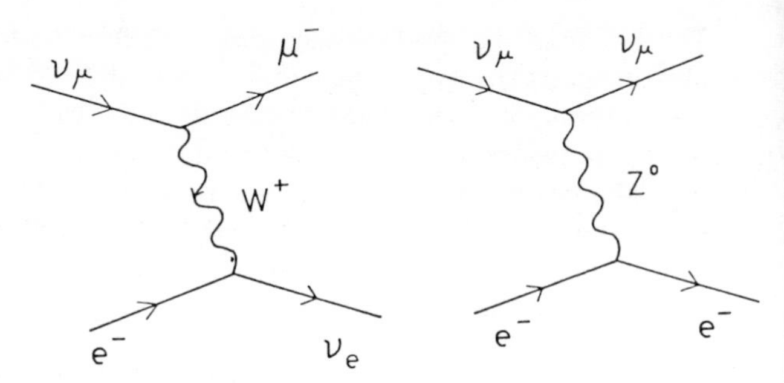

Fig. 8.3 (a) The weak charged-current process $\nu_\mu + e^- \to \mu^- + \nu_e$ and (b) the weak neutral current process $\nu_\mu + e^- \to \nu_\mu + e^-$.

$$\langle \tfrac{1}{2}, +\tfrac{1}{2}|t_+|\tfrac{1}{2}, -\tfrac{1}{2}\rangle = [1 \quad 0] \begin{bmatrix} 0 & 1 \\ 0 & 0 \end{bmatrix} \begin{bmatrix} 0 \\ 1 \end{bmatrix} = 1,$$

so that the μ-lepton current becomes

$$(g/2\sqrt{2})\bar{u}(\mu)\gamma_\mu(1-\gamma_5)u(\nu) = (g/2\sqrt{2})j_\mu(\mu)$$

with a similar expression for the electron current. The complete amplitude for the reaction is

$$T = (g^2/8)j_\mu(\mu)\frac{1}{q^2 - M_W^2}j_\mu(e)$$

where q^2 is the four-momentum of the exchanged virtual W-boson. If this is compared with the Fermi amplitude then

$$G = g^2/(4\sqrt{2}M_W^2). \qquad [8.3]$$

We collect here the other useful expressions

$$e = g \sin\theta_W \qquad [8.1]$$

$$M_Z = M_W/\cos\theta_W. \qquad [8.2]$$

These three expressions relate g, M_W, M_Z and θ_W to G and e but do not fix their absolute values. Table 8.1 summarizes the electroweak charges of the fermions. The terms in brackets denote the coupling to $W^\pm$ or Z^0 available to left-handed fermions whilst those terms not bracketed apply equally for right- or left-handed fermions. The weak currents take the form $\bar{u}\gamma_\mu(1-\gamma_5)u$ for left-handed and $\bar{u}\gamma_\mu(1+\gamma_5)u$ for right-handed fermions; the electromagnetic current is pure vector $\bar{u}\gamma_\mu u$.

Table 8.1 *Electroweak charges*

Coupling to	γ	$W^\pm$	Z^0
Overall factor appearing in current	e	$g/\sqrt{2} = e/\sqrt{2}\sin\theta_W$	$g/\cos\theta_W = e/\cos\theta_W\sin\theta_W$
ν_e, ν_μ, ν_τ	0	$(+\frac{1}{2})$	$(+\frac{1}{2})$
e^-, μ^-, τ^-	-1	$(-\frac{1}{2})$	$(-\frac{1}{2})+\sin^2\theta_W$
u, c, t	$+\frac{2}{3}$	$(+\frac{1}{2})$	$(+\frac{1}{2})-\frac{2}{3}\sin^2\theta_W$
d′, s′, b′	$-\frac{1}{3}$	$(-\frac{1}{2})$	$(-\frac{1}{2})+\frac{1}{3}\sin^2\theta_W$

8.3 The discovery of the weak neutral current

This discovery was made at CERN by a team of physicists using a large bubble chamber filled with Freon (CF_3Br) and exposed to beams of neutrinos or antineutrinos. The beam of neutrinos was generated by allowing the internal proton beam from the CERN proton synchrotron (PS) to fall on a metal target; the emerging positive particles, mostly π^+-mesons, were then collimated forward and the negatives dispersed by a pulsed magnet (horn). After traversing a drift region where the π-mesons decay to μ^+-leptons and neutrinos (ν_μ) the beam encountered a thick wall of concrete and earth. This absorbed any residual hadrons and also stopped the muons so that their decay products had low momentum and could be neglected. The neutrinos had an average energy of between 1 and 2 GeV. Alternatively a beam of antineutrinos could be generated by reversing the polarity of the horn. Gargamelle, the bubble chamber in question, had a useful volume of 6·2 m^3 with an applied magnetic field of 2 T and heavy shielding all around.

Any tracks seen in the bubble chamber could only be

products from neutrino interactions. Charged-current processes

$$\nu_\mu + \text{nucleus} \to \mu^- + \text{hadrons}$$
$$\bar{\nu}_\mu + \text{nucleus} \to \mu^+ + \text{hadrons}$$

show a long, penetrating muon track accompanied by none or one or more hadron tracks which soon interact and get absorbed in the heavy liquid. Track curvature identified the sign and energy of the μ-lepton. The subject of the experimenters' interest was, however, the neutral-current interactions

$$\overset{(-)}{\nu}_\mu + \text{nucleus} \to \overset{(-)}{\nu}_\mu + \text{hadrons} \qquad [8.4]$$

or

$$\overset{(-)}{\nu}_\mu + e^- \to \overset{(-)}{\nu}_\mu + e^- \qquad [8.5]$$

In the former case only hadrons would be observed to emerge from the interaction, and in the latter a lone electron. The experimenters' first success was to report finding (among 735,000 pictures) the example of scattering from an electron which is shown in Fig. 8.4 (Hasert (1973a)). The recoiling electron produces photons by bremsstrahlung and these in turn convert to electron-positron pairs. Such features make the identification of an electron track very secure. Subsequently the same authors (Hasert (1973b)) reported the observation of examples of the neutrino-induced reaction [8.4] and its antineutrino-induced counterpart. The weak neutral coupling is

$$\gamma_\mu (g/\cos\theta_W)(t_3^L - Q\sin^2\theta_W).$$

The dependence of the weak neutral coupling on $\sin^2\theta_W$ is extremely valuable because it makes it possible to determine the Weinberg mixing angle from measurements of weak-current cross-sections. As an example we take the reaction

$$\nu_\mu + e^- \to \nu_\mu + e^-,$$

where both currents are neutral. The procedure to be followed for calculating the cross-section has been given in Chapter 6 and Appendix E; it yields

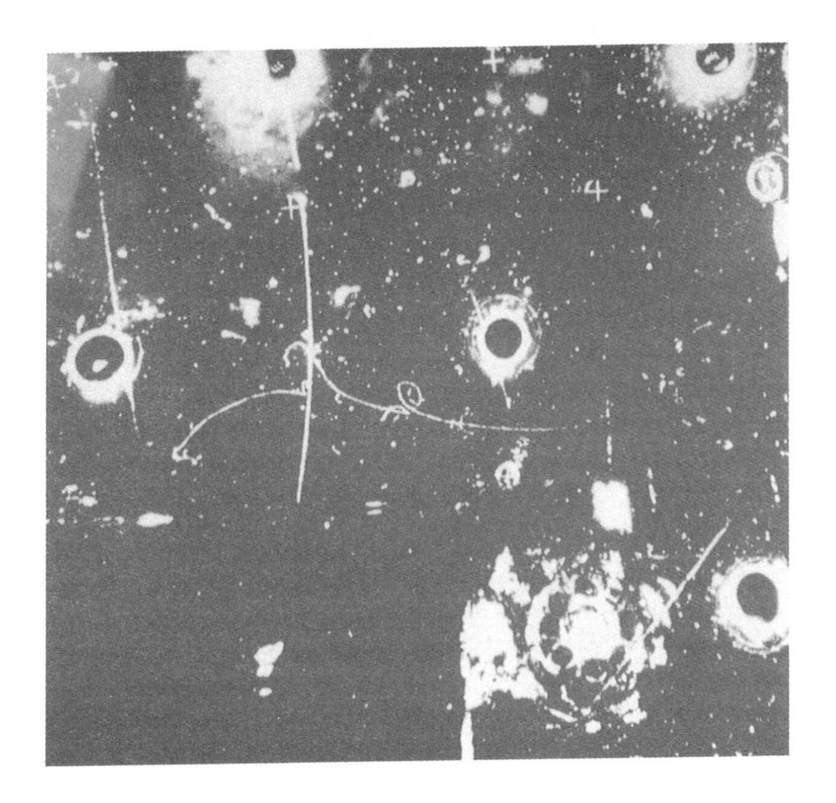

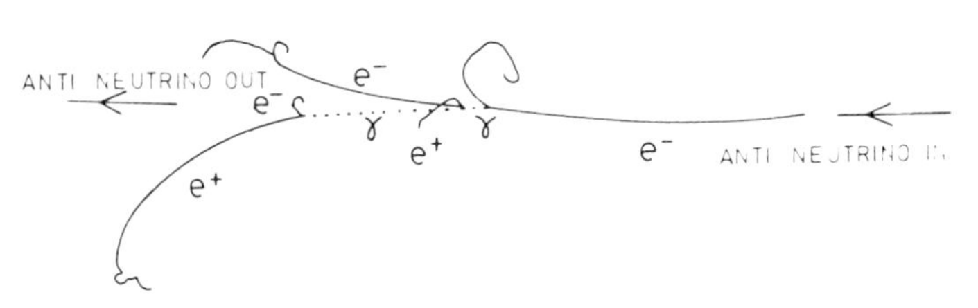

Fig. 8.4 Photograph of the first weak neutral-current event $\bar{\nu}_\mu + e^- \rightarrow \bar{\nu}_\mu + e^-$ recorded in the Gargamelle Bubble Chamber (Hasert, 1973a). (Photograph courtesy CERN.) The annotated line diagram shows the identities of particles seen and unseen (γ and $\bar{\nu}$).

$$\sigma(\nu_\mu e^- \rightarrow \nu_\mu e^-) = (G^2 s/4\pi)[3 - 12 \sin^2 \theta_W + 16 \sin^4 \theta_W].$$

A similar analysis gives the antineutrino cross-section:

$$\sigma(\bar{\nu}_\mu e^- \rightarrow \bar{\nu}_\mu e^-) = (G^2 s/4\pi)[1 - 4 \sin^2 \theta_W + 16 \sin^4 \theta_W]$$

(where s is equal to $2E_\nu m_p$).

Measurements have been made of both these cross-sections with electronic detectors and Bergsma (1983) finds

$$\sigma(\nu_\mu e) = (2{\cdot}1 \pm 0{\cdot}55 \pm 0{\cdot}49) \times 10^{-42}\ E_\nu\ \text{cm}^2\ \text{GeV}$$
$$\sigma(\bar{\nu}_\mu e) = (1{\cdot}6 \pm 0{\cdot}35 \pm 0{\cdot}36) \times 10^{-42}\ E_\nu\ \text{cm}^2\ \text{GeV}$$

giving $\sin^2 \theta_W = 0{\cdot}215 \pm 0{\cdot}04 \pm 0{\cdot}015$, where in each case the first error is statistical and the second is systematic. Experiments on neutrinos scattering from quarks inside nucleons and on electroweak interference have also been used to measure $\sin^2 \theta_W$. Interference effects are seen in the angular distribution for

$$e^+ + e^- \rightarrow l^+ + l^-$$

where l is e, μ or τ. The initial lepton pair can annihilate to give either a photon or a Z^0 and the interference of these two amplitudes leads to an asymmetric angular distribution in $\cos \theta$. Another interference effect is the appearance of small parity-violation terms in atomic decays. Measurements of $\sin^2 \theta_W$ by alternative techniques are consistent and give (Kim, 1981)

$$\sin^2 \theta_W = 0{\cdot}233 \pm 0{\cdot}009.$$

The determination of $\sin^2 \theta_W$ makes it possible to extract values for the masses of the $W^\pm$ and Z^0 bosons from equations [8.1], [8.2] and [8.3]. Inserting the values of $\sin^2 \theta_W$ directly gives a W mass of 77 GeV/c^2. However, at this large mass, α the fine-structure constant has changed appreciably from its value measured at low energy due to renormalization effects. There are also radiative corrections due to the self-energy of the W and Z^0 bosons and to the value of G extracted from muon-decay experiments. All these radiative corrections can be absorbed into one term written $1 - \Delta r$ where Δr is approximately 0·07.

Then the corrected value for the W-boson mass is

$$M_W = [e^2/4\sqrt{2}\, G \sin^2 \theta_W (1 - \Delta r)]^{\frac{1}{2}}$$
$$= 38{\cdot}65/\sin \theta_W$$
$$= 80{\cdot}0 \pm 3 \text{ GeV}/c^2$$

and

$$M_Z = 91{\cdot}3 \pm 2{\cdot}5 \text{ GeV}/c^2.$$

The weak vector bosons are predicted to be a hundred times as heavy as the proton.

8.4 The discovery of the $W^{\pm}$ and Z^0 bosons

The production of these bosons requires considerable energy. From the collision of a proton of energy E with a target proton at rest the available CM energy is only $\sqrt{2Em_p}$ (23 GeV for E equal to 270 GeV). Rubbia, McIntyre and Cline (1977) proposed using the CERN Super Proton Synchrotron (SPS) to accelerate simultaneously counter-rotating proton and antiproton beams. When these beams are brought into collision the available CM energy rises to $2E$, enough to produce the weak vector bosons. The scheme worked thanks to methods introduced by Van der Meer (1972) to collect antiprotons in sufficient quantities. The electroweak theory predicts the rate at which bosons will be produced, namely one $W^{\pm}$ per 10^7 $\bar{p}p$ collisions and one Z^0 per 10^8 collisions at a beam energy of 270 GeV. Examples of the production processes are

$$\bar{u} + d \rightarrow W^-$$
$$\bar{u} + u \rightarrow Z^0,$$

where the (anti)quark originates from the (anti)proton. These bosons have lifetimes of only 10^{-23} s because of the large energy release; therefore only their decay products can be observed. The vector bosons decay to quarks and to leptons

$$W^- \rightarrow \bar{u} + d \quad \text{or} \quad e^- + \bar{\nu}_e,$$

where the rate for the quark channel is enhanced by a colour factor of 3. With three generations of quarks and leptons available the branching ratio for $(e^- + \bar{\nu}_e)$ is $\frac{1}{12}$. Decays to quarks are followed by quark fragmentation into jets of hadrons; these decays are swamped by the background of jets from QCD processes. However, the leptonic decays

$$W^- \to e^- + \bar{\nu}_e \quad \text{or} \quad \mu^- + \bar{\nu}_\mu$$
$$Z^0 \to e^+ + e^- \quad \text{or} \quad \mu^+ + \mu^-$$

have negligible backgrounds from other processes. If bosons are produced at rest in the laboratory frame the decay leptons travel in opposite directions, each with energy of approximately $M_W/2$ (or $M_Z/2$). Two sets of apparatus were built which had the primary aim of being effective detectors of 40 GeV leptons.

Figure 8.5 shows the UA1 detector installed at one $\bar{p}p$ intersection point. At the centre is a tracking chamber which records charged-particle trajectories. This is immersed in a 0·7 tesla magnetic field so that measurements of track curvature can be used to determine particle momenta. Electron identification requires a track pointing to an electromagnetic shower in the electromagnetic calorimeter and a match between this shower energy and the track momentum as determined from curvature. Muon identification relies on penetration, through the layers of iron and lead, to drift chambers external to the apparatus. The search for the Z^0 is the easier to describe. A study was made of all interactions emitting a pair of oppositely charged leptons. Suppose the leptons have four-momenta $(E_1, \mathbf{p}_1)$ and $(E_2, \mathbf{p}_2)$. Then their parent (if they come from a decay) has four-momentum $(E_1 + E_2, \mathbf{p}_1 + \mathbf{p}_2)$. Its mass is given by

$$M^2 = (E_1 + E_2)^2 - (\mathbf{p}_1 + \mathbf{p}_2)^2.$$

Figure 8.6 shows the mass spectrum of opposite-sign lepton pairs collected between 1982 and 1984. It consists of a narrow peak near 90 GeV/c^2 separated by a gap of 30 GeV/c^2 from the rapidly falling tail of the background. This peak is interpreted as the product of Z^0 decays. The mass of the Z^0 obtained from a weighted average of these events is

$$M_Z = 93{\cdot}0 \pm 3{\cdot}5 \text{ GeV}.$$

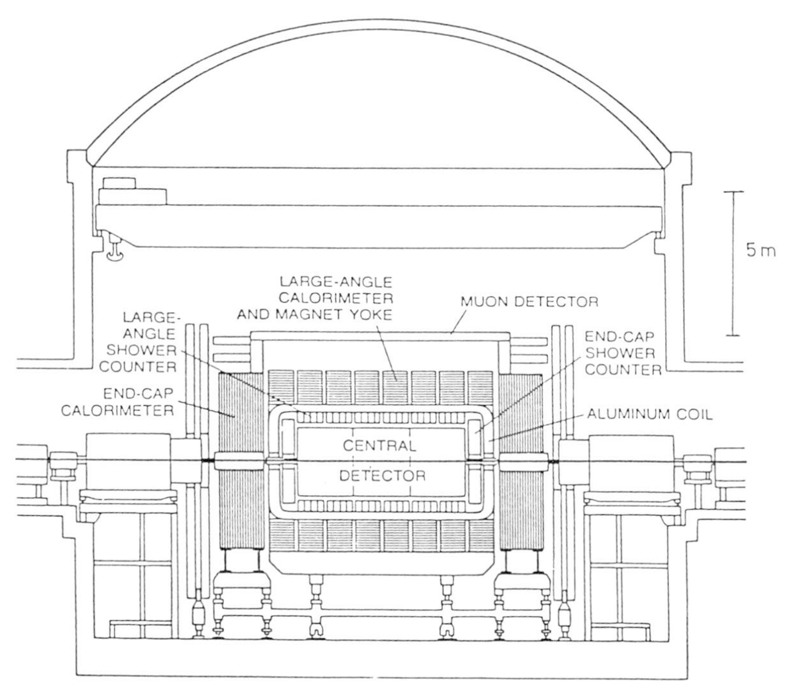

Fig. 8.5 The UA1 experiment; a vertical section containing the beam axis is shown. (Photograph courtesy *Scientific American*, March 1982.)

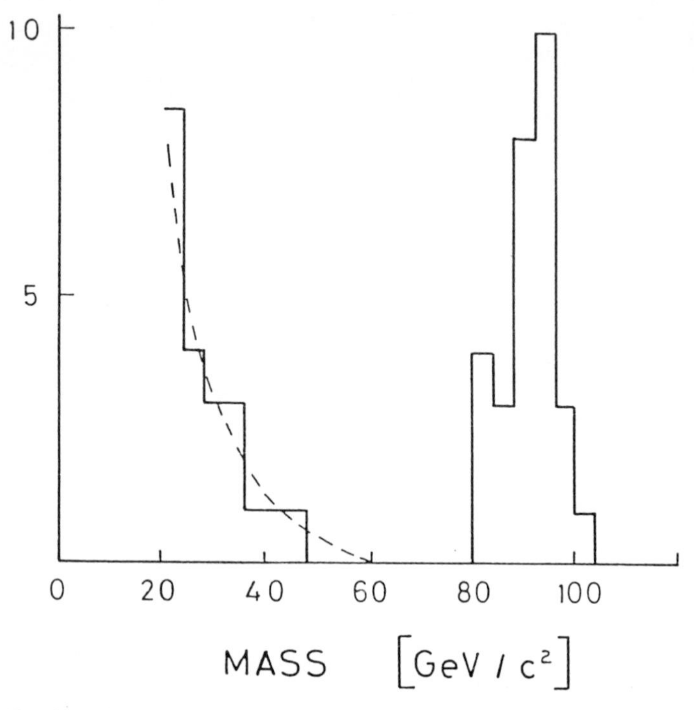

Fig. 8.6 The effective mass spectrum of unlike-sign dilepton pairs ($\mu^+\mu^-$, e^+e^-) detected in the UA1 apparatus.

The frequency of W-decays (see Fig. 3.9) was around ten times greater, as expected, and gave

$$M_W = 83{\cdot}5 \pm 3{\cdot}0 \text{ GeV}.$$

Both mass values are in excellent agreement with the predictions of electroweak theory. Besides having the correct masses and production rates, it is also observed that the W-bosons show charge asymmetry (i.e. C and P violation) in their decays. The experimental angular distribution of the electrons from W^--decays follows the highly asymmetric form predicted by the $V-A$ theory, $(1+\cos\theta)^2$, where θ is the polar angle the electron direction makes with the incident proton direction. The angular distribution of positrons from W^+-decay is the mirror image

$(1-\cos\theta)^2$. It is quite easy to see why electrons can be emitted at $\theta=0$ but not at $\theta=180°$. Only the left-handed (right-handed) component of the d-quark (ū-antiquark) takes part in the interaction. Schematically:

$$\bar{\mathrm{p}}\ \underrightarrow{\text{direction}}\qquad \underset{\rightarrow}{\bar{\mathrm{u}}}\ \underset{\rightarrow}{\mathrm{d}}\qquad \mathrm{p}\ \underleftarrow{\text{direction}}.$$

The spatial part of the wavefunction of the incoming quark (or antiquark) is a plane wave

$$\psi=\exp[i(pz-Et)],$$

whose component of orbital angular momentum along the beam axis $0z$ is given by

$$L_z\psi\equiv -i(x\partial/\partial y-y\partial/\partial x)\psi=0.$$

Conservation of the z-component of total angular momentum (J_z) therefore requires that the W^- produced from ūd annihilation has its unit spin thus:

$$\underset{\rightarrow}{\mathrm{W}}{}^{-}.$$

In collinear decays ($\theta=0$ or $180°$) L_z again vanishes. J_z will only be conserved in the collinear decays if the daughter electron and antineutrino have their spins aligned like this:

$$\mathrm{e}^-\ \underleftarrow{\text{direction}}\qquad \underset{\rightarrow}{\mathrm{e}^-}\ \ \underset{\rightarrow}{\bar{\nu}}\qquad \bar{\nu}\ \underrightarrow{\text{direction}}.$$

Consequently the antineutrino, being right-handed, can only travel right-ward, and the electron goes left along the incident proton direction: $\theta=0$ is allowed but $\theta=180°$ is forbidden.

The width of the Z^0 is of special interest because it provides a way of counting the number of (light) neutrino species. Each decay of the Z^0 to a given neutrino species

$$Z^0\rightarrow\nu_l+\bar{\nu}_l$$

contributes 0·18 GeV/c^2 to its natural width. With the three observed generations of leptons and quarks a width of 2·94 GeV/c^2 is to be expected. A direct measurement of Γ_z should be possible either from antiproton-proton collider experiments or from measurements of Z^0 production at the future large e^+e^- colliders being built at CERN (LEP) and at Stanford (SLC).

8.5 Strangeness-changing neutral currents (SCNC)

The coupling of s′-quarks to the Z^0-boson generates strangeness-changing components; symbolically

$$\begin{aligned} Z^0 s'\bar{s}' &= Z^0(s \cos\theta_C - d \sin\theta_C)(\bar{s}\cos\theta_C - \bar{d}\sin\theta_C) \\ &= Z^0(s\bar{s}\cos^2\theta_C + d\bar{d}\sin^2\theta_C - (s\bar{d} + d\bar{s}) \\ &\quad \cos\theta_C \sin\theta_C), \end{aligned}$$

where θ_C is the Cabibbo angle. However, the corresponding coupling to d′-quarks gives matching components

$$\begin{aligned} Z^0 d'\bar{d}' &= Z^0(s \sin\theta_C + \mathrm{d} \cos\theta_C)(\bar{s}\sin\theta_C + \bar{d}\cos\theta_C) \\ &= Z^0(s\bar{s}\sin^2\theta_C + d\bar{d}\cos^2\theta_C + (s\bar{d} + d\bar{s}) \\ &\quad \cos\theta_C \sin\theta_C). \end{aligned}$$

Summing we obtain:

$$Z^0(s'\bar{s}' + d'\bar{d}') = Z^0(s\bar{s} + d\bar{d}).$$

The Z^0 exchange does not therefore contribute to SCNC. The argument can be extended to include the coupling to the b-quark. Then it follows that

$$Z^0(d'd' + s'\bar{s}' + b'\bar{b}') = Z^0(\bar{d}d + s\bar{s} + b\bar{b})$$

because the Kobayashi-Maskawa mixing matrix is unitary. All *flavour*-changing neutral currents are therefore excluded for the Z^0 coupling. However, K^0_L-decay to leptons

$$K^0_L \to \mu^+ + \mu^-$$

can occur through the box diagrams shown in Fig. 8.7. This SCNC process has a branching fraction measured to be only $(9{\cdot}1 \pm 1{\cdot}9) \times 10^{-9}$. The amplitudes for the two box diagrams contain the factors $-g^4 \cos\theta_C \sin\theta_C$ and $+g^4 \cos\theta_C \sin\theta_C$ respectively and would cancel in the limit that the u- and c-quarks had equal masses. Nature, it seems, has carefully

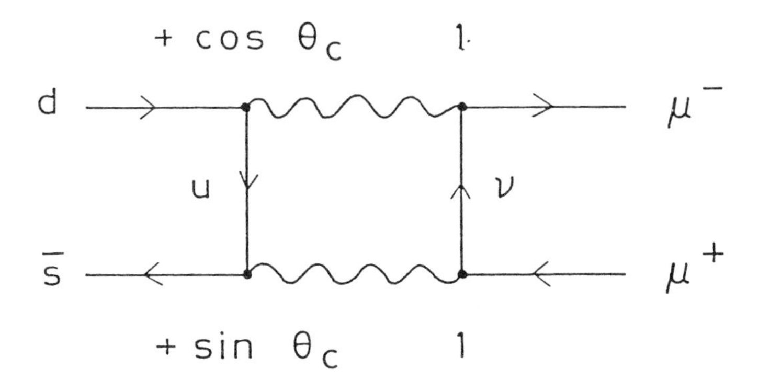

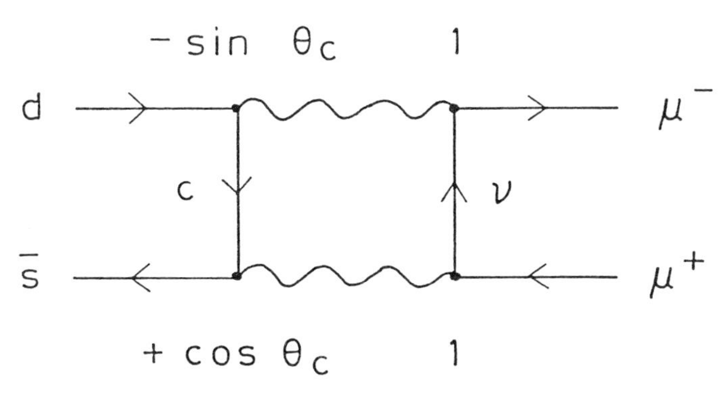

Fig. 8.7 Box diagrams for the decay $K_L^0 \rightarrow \mu^+ + \mu^-$, with the Cabibbo factors entered against each vertex.

organized the suppression of SCNC effects. The cancellation achieved between the box diagrams was first noticed by Glashow, Iliopoulos and Maiani (1970) and it led them to propose the existence of the charm quark in order to explain the low rate of SCNC processes. This conjecture provided the original impetus in the search for charm.

8.6 Particle number conservation laws

We have seen how the application of local gauge invariance has led to successful models of the strong and electroweak interactions. The coupling of fermions to the field bosons takes the diagrammatic form shown in Fig. 4.1 and was introduced first for electromagnetism. At a vertex involving a given lepton species the net number of leptons is not changed. This fact underlies the conservation laws for leptons noted in Chapter 2. Equally the net number of quarks is not changed at any strong or electroweak vertex. However, because quarks are only present in nature bound into hadrons this observation translates into a conservation law for the net number of baryons (qqq structures). Mesons ($\bar{q}q$ structures) can be created or destroyed in any number provided energy is conserved.

Chapter 9
Electroweak probes of hadron structure

The gauge theories of the electroweak and strong force form the 'standard model' based on $SU(3) \otimes SU(2)_L \otimes U(1)$ symmetry. We have seen its successes: the unified electroweak theory incorporates the faithful description of electromagnetic phenomena found in QED and predicts the existence of the weak neutral currents and the masses of the gauge bosons $W^{\pm}$ and Z^0; for the strong interactions QCD reconciles the asymptotic freedom of quarks at short range with their confinement in hadrons. It is now appropriate to move on to discuss the structure and properties of those more complex elementary particles, the hadrons. These properties cannot be calculated directly from QCD in most cases because of unresolved difficulties in the non-perturbative regime. In this chapter the use of deep inelastic lepton scattering to probe hadron structure forms the main topic.

The primary result of deep inelastic lepton scattering from nucleons was introduced in Chapter 2 (see Fig. 2.4): the angular distribution of scattering is of the form expected if there are point-like constituents within the nucleon. Experiments using muon and neutrino beams have confirmed the original result obtained with electron beams. When a lepton scatters inelastically (or elastically) from a hadron the small value of the electroweak coupling constant ensures that the interaction is dominantly via the exchange of a single boson (γ, W or Z^0). Double-photon exchange is less important by a factor α^2 and so on. The methods for calculating cross-sections for single-boson exchange processes have already been fully developed in Chapters 4 and 7 and will be employed here. 'Deep' inelastic implies that the four-momentum transfer squared $(-q^2)$ is large compared to the scale (Λ^2) of strong interactions between quarks. A deep inelastic scattering therefore involves exchange of a single electroweak boson between the beam lepton and a quark (or antiquark) which is effectively isolated from its fellows in the target nucleon. Thus the experiments can measure the momentum distributions of individual quarks inside nucleons. The quarks (plus antiquarks) are found to account for only about 50 per cent of the nucleon momentum: the remainder of the momentum must be assigned to the gluons which are invisible to the electroweak probes. Mesons are not available as stable targets and a related kind of experiment making use of the Drell-Yan process is required to explore their internal structure. This purely electromagnetic process proceeds in two steps: first a quark from one hadron annihilates on an antiquark from another hadron to give a virtual photon; subsequently the virtual photon converts to a lepton-antilepton pair which is observed experimentally.

9.1 Deep inelastic electron scattering

The underlying process involving exchange of a single virtual photon between lepton and nucleon is drawn in Fig. 9.1. $k(k')$ is the four-momentum of the incident (outgoing) lepton, P the four-momentum of the target nucleon and q the four-momentum transfer. θ is the scattering angle of the lepton measured in the laboratory. We also define $Q^2 \equiv -q^2$. Initially the lepton scatters elastically from a quark and subsequently the nucleon shatters into a number of hadrons. This impulse model with a single active quark and the remainder acting as spectators is justified in QCD if the four-momentum transfer squared (Q^2) from lepton to quark is very much larger than Λ^2. Put another way: if the time of interaction $1/\sqrt{Q^2}$ is short compared to the time scale Λ^{-1} of parton interactions then the remaining partons can be treated as spectators. In practice Q^2 needs to be bigger than $1(\text{GeV}/c)^2$. Experiments with incident electron or muon beams make use of a single-arm spectrometer which measures both the direction and energy of the scattered

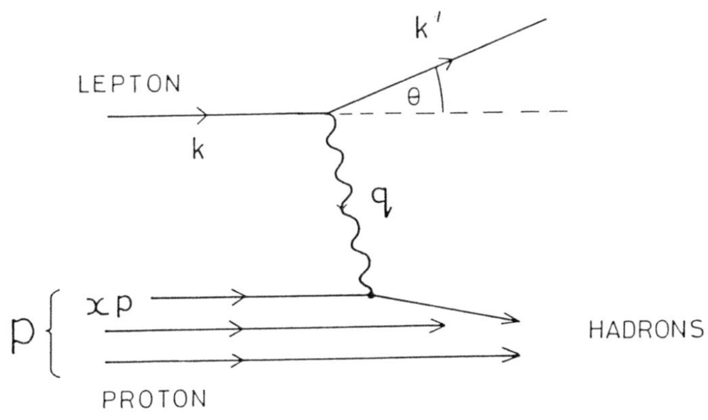

Fig. 9.1 The reaction underlying deep inelastic lepton scattering: a lepton-quark elastic scatter. $k_\mu(k'_\mu)$ is the four-momentum of the incident (outgoing) lepton, $xP_\mu(P_\mu)$ that of the incident quark (nucleon) and q_μ that carried by the exchanged vector boson.

lepton in the laboratory. The results can be expressed using the double differential cross-section $d^2\sigma/d\Omega\, dE'$, where E' is the energy of the recoil lepton and $d\Omega$ is the solid angle element around the direction θ in the laboratory frame. The useful Lorentz scalars are now listed (some with their explicit values in the laboratory frame):

q^2 the four-momentum transfer squared carried by the photon exchanged,

ν the energy transfer $(E-E')$,

$x=-q^2/2M\nu$,

$y=\nu/E$.

M is the target nucleon mass. Definitions of x and y valid in any inertial frame are these:

$$x=-q^2/2P\cdot q$$
$$y=P\cdot q/P\cdot k$$

where P, q and k are respectively the four-momenta of the target nucleon, the exchanged boson and the incident lepton (Fig. 9.1). Note that these dot products are four-vector dot products; so, for example, $p\cdot k\equiv p_\mu k_\mu=p_0k_0-\mathbf{p}\cdot\mathbf{k}$. x bears a simple physical interpretation relative to the active quark. x is the fraction of the nucleon four-momentum carried by this quark. In support of this assertion we note first that the four-momenta of this quark before and after absorbing the boson are xP_μ and $xP_\mu + q_\mu$ respectively. The quark mass remains unchanged, so that we have

$$(xP)^2 = (xP + q)^2$$

i.e.

$$x^2P^2 = x^2P^2 + 2xP \cdot q + q^2$$

and therefore

$$x = -q^2/2P \cdot q,$$

which matches the definition given above. The interpretation of x as a momentum fraction was originally made for frames in which the nucleon has infinite momentum. Then x can range from *zero* to *unity* only. This view of x is quite tenable in the overall CM frame when the beam energy is substantial. Note also that the transverse momentum components which result from fermi motion inside the nucleon are then relatively small.

The cross-section measured in deep inelastic lepton scattering is an inclusive cross-section: i.e. only the scattered lepton is detected and there may be any number of hadrons emerging and any kinematically allowed partition of energy amongst these hadrons. This makes the calculation of the cross-section easy. It is sufficient to calculate the cross-section for the first stage, namely the lepton-quark elastic scattering; the probability for the rearrangement of the recoiling quark and its spectator colleagues into outgoing hadrons is simply unity. The cross-section for electron-quark elastic scattering can be written down using results given in Appendix E. The cross-section is

$$d\sigma = A \, dQ_2/F,$$

where dQ_2 is two-body phase space for the outgoing lepton plus quark and F is the flux factor. In the laboratory frame:

$$A = (8e^4Q_f^2/q^4)[(k' \cdot p')(k \cdot p) + (k' \cdot p)(k \cdot p') - M^2x^2(k' \cdot k)],$$

where Mx is the portion of target nucleon mass assigned to the active quark and $Q_f e$ is its charge, while $p = xP$.

When the electron mass is neglected these expressions lead to a differential cross-section

$$\mathrm{d}\sigma/\mathrm{d}\Omega = [4Q_f^2 \alpha^2 (E')^3/Eq^4][\cos^2(\theta/2) + (\nu/Mx)\sin^2(\theta/2)].$$

The cross-section for scattering from the whole nucleon is the incoherent sum of the contributions from all the quarks it contains. $\rho_f(x)\mathrm{d}x$ is defined to be the number of quarks carrying a fraction between x and $x + \mathrm{d}x$ of the parent nucleon momentum, for flavour f. Then the cross-section for the nucleon is

$$\mathrm{d}\sigma/\mathrm{d}\Omega = [4\alpha^2 (E')^3/Eq^4][\cos^2(\theta/2) + (\nu/Mx)\sin^2(\theta/2)] \times \Sigma Q_f^2 \rho_f(x)\mathrm{d}x.$$

Now the choice of x in fact fixes the energy E' of the recoil lepton. It is convenient to express this result in terms of the measured quantities (E', θ) rather than (x, θ). To make the conversion the factor $(\partial x/\partial E')_\theta = x/yE'$ is needed. Then the differential cross-section becomes

$$\mathrm{d}^2\sigma/\mathrm{d}\Omega\,\mathrm{d}E' = [4\alpha^2 (E')^2/q^4][(x/\nu)\cos^2(\theta/2) + (1/M)\sin^2(\theta/2)]\,\Sigma_f \rho_f(x) Q_f^2.$$

This cross-section formula which has been derived using the quark model can be compared with an expression that is valid for a single-photon exchange irrespective of the structure of the nucleon:

$$\mathrm{d}^2\sigma/\mathrm{d}\Omega\,\mathrm{d}E' = [4\alpha^2 (E')^2/q^4][W_2(\nu, q^2)\cos^2(\theta/2) + 2W_1(\nu, q^2)\sin^2(\theta/2)],$$

in which W_1 and W_2 are called structure functions. When the coefficients of $\sin^2 \theta/2$ and $\cos^2 \theta/2$ in the last two expressions are compared, we find that

$$F_1 = 2MW_1 = \Sigma_f\, \rho_f(x) Q_f^2$$
$$F_2 = \nu W_2 = \Sigma_f x \rho_f(x) Q_f^2.$$

We see that according to the quark model the structure functions are simply the quark longitudinal momentum distributions. In general the structure functions would be functions of the two independent kinematic variables ν and q^2; whilst in the quark model the structure functions depend only on the combination $-q^2/2M\nu$, i.e. on x. Such behaviour of the structure functions is termed scaling and paradoxically it indicates that the constituents of the nucleon do not provide any scale; they are point-like. There is a further strict relationship between the structure functions if the constituents of the nucleon have spin $\frac{1}{2}$, namely

$$xF_1 = F_2,$$

which is called the Callan-Gross relation. The quark-model predictions satisfy this relation.

Comparisons made between structure functions measured at identical values of x for a range of q^2 values are shown in Fig. 9.2. The measured structure functions are seen to vary only weakly (logarithmically) with $-q^2$. Violation of the Callan-Gross relation is at the few per cent level. These departures from precise scaling are small enough that the simple quark-model predictions retain considerable usefulness. The weak scaling violations are explained in QCD when account is taken of the emission of gluons by the quarks: this feature will be discussed later.

The quantities of physical interest which emerge from deep inelastic scattering experiments are the quark longitudinal momentum distributions $\rho_\alpha(x)$ for a quark of species α. For convenience let us write $u(x)$ for $\rho_u(x)$ and $\bar{u}(x)$ for $\rho_{\bar{u}}(x)$ inside a proton. Then for a proton target:

$$F_2(\text{ep}) = x[(4/9)(u+\bar{u}) + (1/9)(d+\bar{d})].$$

Heavy flavour contributions are not very important because a lot of energy is needed to create for example, a $b\bar{b}$ pair. The quark content of the neutron is obtained from the proton by making the interchange u$\leftrightarrow$d, so that for a neutron target:

$$F_2(\text{en}) = x[(4/9)(d+\bar{d}) + (1/9)(u+\bar{u})].$$

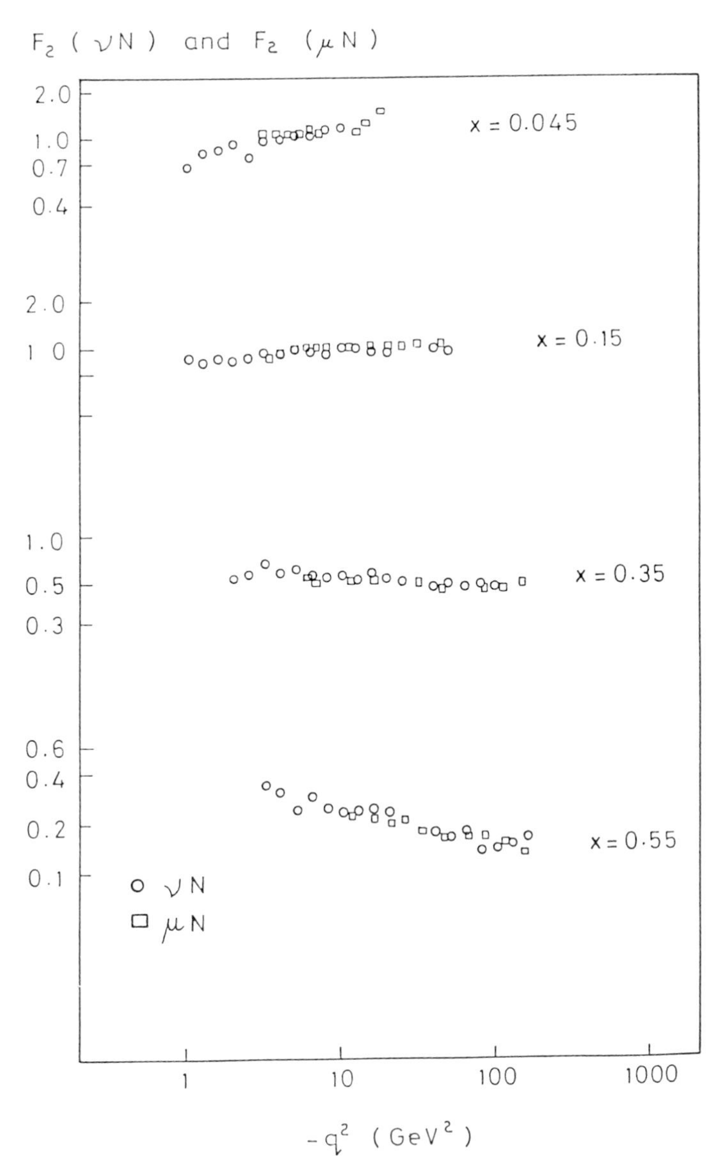

Fig. 9.2 The structure function $F_2(x)$ measured with muon and neutrino beams incident on iron targets. Only representative values of x are shown. (After Demioz, Ferroni and Longo, *Phys. Rep.* **130**, 293 (1986)).

Finally, for an isoscalar target (equal numbers of protons and neutrons) such as ^{12}C, ^{16}O or ^{4}He:

$$F_2(\mathrm{eN}) = (5/18)x[u + \bar{u} + d + \bar{d}].$$

9.2 Deep inelastic neutrino scattering

Neutrino scattering provides complementary information on the nucleon structure which is not directly accessible with charged leptons. The underlying 'elastic' processes are

$$\nu_\mu + \mathrm{d} \to \mu^- + \mathrm{u} \qquad [9.1]$$

$$\nu_\mu + \bar{\mathrm{u}} \to \mu^- + \bar{\mathrm{d}} \qquad [9.2]$$

and the charge conjugate reactions for antineutrinos. In these cases a W-boson is exchanged. The differential cross-section for fermion-fermion scattering due to the weak force is given in Appendix E.

$$\mathrm{d}\sigma/\mathrm{d}\Omega^* = (G^2 s/4\pi^2)$$

for reaction [9.1] in the neutrino-quark CM frame is isotropic; and for reaction [9.2]

$$\mathrm{d}\sigma/\mathrm{d}\Omega^* = (G^2 s/16\pi^2)(1 + \cos\theta^*)^2,$$

which vanishes in the backward direction. (Note that s is the CM energy squared for the neutrino-quark system.) In both reactions the orbital angular momentum is zero because the interaction is a point-contact interaction. In reaction [9.1] all the particles are left-handed, so the net spin component and hence of **J**, along the incident (or outgoing) direction in the CM frame is zero ($J_z = 0$). Emission at any angle is equally likely. On the other hand for reaction [9.2] the net spin component along the direction of the incoming ν_μ is $J_z = -1$, because the target is right-handed. In the final state the spin components are again aligned and so backward scattering would give $J_z = +1$ and is not possible. This feature shows why the quoted angular distribution falls to zero at $\cos\theta^* = -1$. A repetition of the analysis shows that the angular distributions for $\bar{\nu}\mathrm{q}$ is identical to that for $\nu\bar{\mathrm{q}}$ scattering. Similarly $\nu\mathrm{q}$ and $\overline{\nu\mathrm{q}}$ give identical angular distributions. Now we revert to the variables relevant in deep inelastic scattering, namely x and y.

$$y = P \cdot q/P \cdot k = p \cdot q/p \cdot k,$$

where $p = xP$ is the quark four-momentum. Therefore

$$y = p \cdot (k - k')/p \cdot k.$$

Evaluating these terms in the quark-neutrino CM frame, where all the energies and three-momenta are k^* and the scattering angle is θ^* (see Appendix A):

$$p \cdot k = 2k^{*2}$$

$$p \cdot k' = (1 + \cos\theta^*)k^{*2}$$

$$\therefore \; y = (1 - \cos\theta^*)/2$$

$$\therefore \; \mathrm{d}\Omega^* = 2\pi\, \mathrm{d}(\cos\theta^*)$$

$$= 4\pi\, \mathrm{d}y.$$

Also s, the CM energy squared for the neutrino-quark system, is given by

$$s = 2xME,$$

where xM is the equivalent mass of the target quark and E is the laboratory energy of the incident neutrino. In the case of νq or $\overline{\nu \mathrm{q}}$ scattering

$$\mathrm{d}\sigma/\mathrm{d}y = (2G^2 xME/\pi),$$

while for $\nu\bar{\mathrm{q}}$ or $\bar{\nu}$q scattering

$$\mathrm{d}\sigma/\mathrm{d}y = (2G^2 xME/\pi)(1 - y)^2.$$

It is now easy to write down the cross-section for deep inelastic neutrino scattering off a proton. This will be the incoherent sum of the contributions from individual quarks and antiquarks:

$$\mathrm{d}^2\sigma(\nu\mathrm{p})/\mathrm{d}y\,\mathrm{d}x = (G^2 ME/\pi)[2xd + 2(1 - y)^2 x\bar{u}],$$

$$\mathrm{d}^2\sigma(\bar{\nu}\mathrm{p})/\mathrm{d}y\,\mathrm{d}x = (G^2 ME/\pi)[2x\bar{d} + 2(1 - y)^2 xu].$$

The corresponding expressions for neutrons can be obtained by interchanging u and d. Then, for an isoscalar target with an equal mixture of neutrons and protons (e.g. ^{12}C),

$$d^2\sigma(\nu N)/dy\,dx = (G^2ME/\pi)[x(u+d)+(1-y)^2x(\bar{u}+\bar{d})],$$
$$d^2\sigma(\bar{\nu} N)/dy\,dx = (G^2ME/\pi)[x(\bar{u}+\bar{d})+(1-y)^2x(u+d)].$$

The difference in the y-dependence of the quark and antiquark contributions makes it easy to separate the quark and antiquark structure functions. A well-determined combination of the structure functions is

$$F_2(\nu N) = x[u+d+\bar{u}+\bar{d}].$$

Integrating over the kinematic range of y (zero to unity) gives

$$d\sigma(\nu N)/dx = (G^2ME/\pi)[x(u+d)+x(\bar{u}+\bar{d})/3]$$
$$d\sigma(\bar{\nu} N)/dx = (G^2ME/\pi)[x(\bar{u}+\bar{d})+x(u+d)/3],$$

which are results we use in the next section.

9.3 The nucleon structure functions

The measurements of $F_2(\nu N)$ and $F_2(\mu N)$ are compared in Fig. 9.2 (Demioz, Ferroni and Longo (1986)) for experiments using iron targets. It is immediately clear that, in accord with the quark-model prediction, the two very different techniques are indeed measuring the same quantity. The observed slow variation of $F_2(x)$ with q^2 is in excellent agreement with QCD predictions. Fig. 9.3 shows schematically the quark and antiquark x-distributions [$xq(x)$ and $x\bar{q}(x)$] extracted from the measurements of Bodek *et al.* (1980) and de Groot *et al.* (1979). The area under the curve between x and $x+dx$ is the fraction of the nucleon longitudinal momentum carried by quarks (or antiquarks) with x values between x and $x+dx$. To extract the number of quarks in the same range the ordinate should be divided by x. It can be seen that the number of quarks and antiquarks increases indefinitely as x approaches zero: which means that there are large numbers of low-x quarks and antiquarks present in addition to the expected three 'valence' quarks of the simple Gell-Mann quark model.

A more complete picture of the nucleon structure takes into account the gluons radiated by quarks as they move inside the

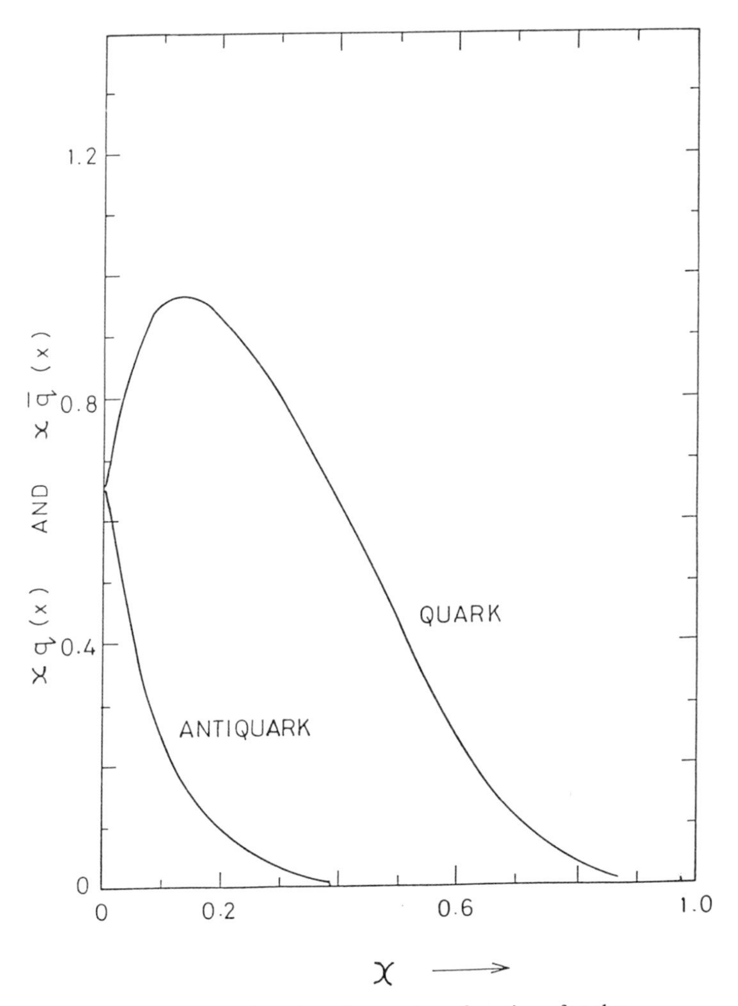

Fig. 9.3 The quark and antiquark structure functions for the nucleon measured with electron and neutrino beams. The four-momentum transfer squared (q^2) range is -4 to -8 (GeV/c)^2.

confining potential provided by their mutual attractions. When a quark is deflected in its motion in the fields of its partners it emits gluon bremsstrahlung. These gluons can in turn convert to quark-antiquark pairs as they move through the fields of the other partons. The sequence continues indefinitely with successive partons in the chain having progressively lower momenta. There is a corresponding increase in α_s as relative momenta decrease, so that we should expect a peaking in the number of quarks and gluons towards x equals zero. This picture is in good agreement with the quark momentum distributions shown in Fig. 9.3. The quark-antiquark pairs generated by gluon conversions are known as the sea; sea quarks and sea antiquarks therefore have identical x-distributions. Splitting up the x-distributions into the 'valence' and 'sea' components

$$q(x)=q_v(x)+q_s(x)$$
$$\bar{q}(x)=\bar{q}_s(x)$$

and $\bar{q}_s(x)=q_s(x)$, which has a common value for light quark species, i.e. $u_s(x)=d_s(x)=s_s(x)$. From Fig. 9.3 we can see that the sea distribution (given by the antiquarks) peaks at x equals zero and the valence contribution $(q(x)-\bar{q}(x))$ has vanished at this point. It is easy to extract from the neutrino cross-sections a value for the net number of quarks in the nucleon, and this provides a check on the consistency of the model. The combination

$$\mathrm{d}\sigma(\nu\mathrm{N})/\mathrm{d}x-\mathrm{d}\sigma(\bar{\nu}\mathrm{N})/\mathrm{d}x=(2G^2ME/3\pi)\,(u+d-\bar{u}-\bar{d}). \quad [9.3]$$

According to the quark model the net number of quarks is

$$\int_0^1 (u+d-\bar{u}-\bar{d})\,\mathrm{d}x\equiv 3,$$

while de Groot *et al.* (1979) find experimentally that the neutrino cross-sections [9.3], when integrated over x, give $3{\cdot}2\pm0{\cdot}5$. Another implication of the quark model as elaborated above is that about half the nucleon momentum should be carried by the gluons which, because they carry neither charge nor weak charge, are invisible to the electroweak probes. An integration of the experimentally determined structure function yields

$$(18/5)\int F_2(\mathrm{eN})\,\mathrm{d}x=\int x(u+\bar{u}+d+\bar{d})\,\mathrm{d}x=0{\cdot}51\pm0{\cdot}08,$$

which indicates that 49 per cent of the nucleon longitudinal momentum is carried by something (the gluons) not sensitive to electroweak probes. Although the gluon x-distribution cannot be measured directly it can be inferred using the quark and antiquark x-distributions and knowledge of the QCD processes $\mathrm{g}\to\mathrm{q}\bar{\mathrm{q}}$ and $\mathrm{q}\to\mathrm{qg}$. Recently Bergsma *et al.* (1983) have summarized the parton momentum distributions measured at values of $-q^2$ from 3 to 50 $(\mathrm{GeV}/c)^2$. These are shown in Fig. 9.4.

Up to this point we have taken the extreme view that the leptons probe the nucleon at sufficiently large four-momentum transfer for the spectator parton interactions with the active quark to be ignored. This would be strictly true if $\alpha_s(q^2)$ were effectively zero. In fact $\alpha_s(q^2)$ is still far from zero at the highest available Q^2 $(\equiv -q^2)$; and Figs. 9.2 and 9.4 reveal that nucleon structure functions show a logarithmic dependence on Q^2. A more complete interpretation of the observed scaling violation starts from the QCD view of a quark (antiquark) in the nucleon being surrounded by a cloud of gluons. At low Q^2 (say Q_1^2) a lepton will probe only the nucleon target on a length scale of Q_1^{-2}; the reader should recall the argument based on the uncertainty principle which is used to connect the momentum transfer and the distance of approach. As Q^2 is increased to Q_2^2 the region probed is reduced and what was seen as a single quark at a scale Q_1^2 is now seen to be a quark plus one or more gluons. This evolution of partons inside partons can be described by what are called the Altarelli-Parisi equations (Altarelli and Parisi (1977)). It is clear immediately from this picture that as $-q^2$ is increased the momentum carried by any parton will in general decrease. The effects of this evolution are that structure functions become more peaked near $x=0$ as $-q^2$ increases, and that the gluons and sea carry increasing fractions

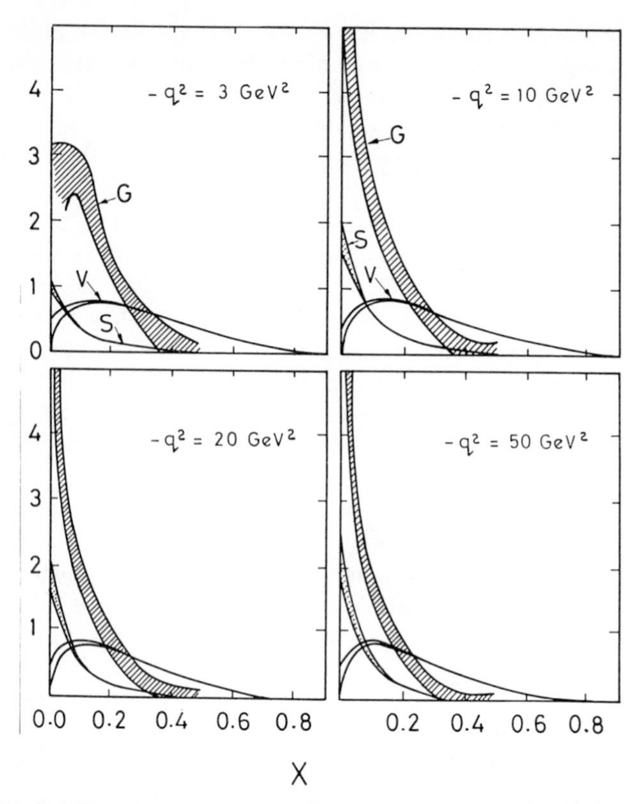

Fig. 9.4 The evolution with $-q^2$ of the x-dependence of the valence quark (V), sea quark (S) and gluon (G) structure functions. The bands express the intervals of uncertainty in the determination of the structure functions (Bergsma *et al. Phys. Lett.* **123B**, 269 (1983)).

of the momentum. These features are nicely brought out by Fig. 9.4.

9.4 The Drell-Yan process

Deep inelastic lepton scattering cannot be used for studying the structure of unstable particles like mesons. This is a nuisance because mesons have a simpler structure than baryons.

Fortunately it is possible to make use of the Drell-Yan process (Drell and Yan (1970)) to explore meson or baryon structure. In this purely electromagnetic process an antiquark from one hadron annihilates on a quark from the other hadron to give a virtual photon, and the photon subsequently converts to a pair of leptons (e^+e^- or $\mu^+\mu^-$). Figure 9.5 shows the process diagrammatically. The remaining partons are regarded as spectators in the initial annihilation; they subsequently re-arrange into a set of final-state hadrons and have little effect on the virtual photon or its daughter leptons. An examination of the kinematics of the process reveals how it is possible to extract information about the parent quark and antiquark. It will be assumed that the CM energy is sufficiently large that masses and transverse momenta of leptons and quarks are negligible in the CM frame. In this same frame let the active quark and antiquark have longitudinal momentum x_1p and $-x_2p$ respectively, where p is the linear momentum of one of the incident hadrons. Let the emerging lepton pair have longitudinal momentum x_Fp. Then

$$x_F = x_1 - x_2$$

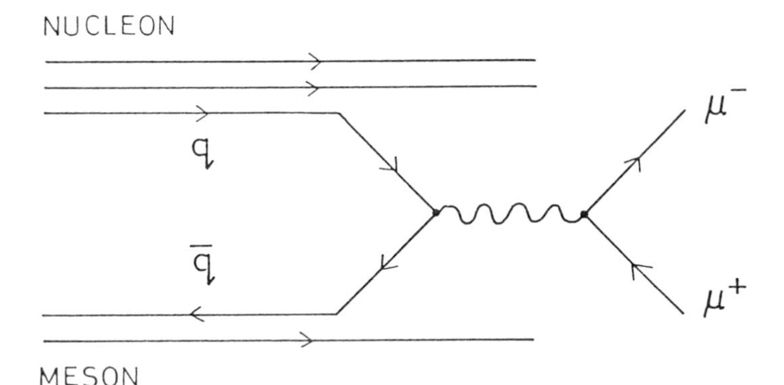

Fig. 9.5 The Drell-Yan process. In this example a valence quark from a nucleon and a valence antiquark from a meson annihilate. Contributions are also made by sea quarks and antiquarks in the hadrons.

and the energy of the pair is

$$E = (x_1 + x_2)p.$$

The effective mass of the lepton pair

$$M^2 = E^2 - x_F^2 p^2 = 4x_1 x_2 p^2.$$

In the limit that the hadron energies greatly exceed their masses the CM energy squared is $s = 4p^2$ and we have the master kinematic expressions for the Drell-Yan process:

$$\tau = M^2/s = x_1 x_2 \quad \text{and} \quad x_F = x_1 - x_2.$$

Measurements of the leptons' four-momenta determine M^2 and x_F and hence fix the x-values of the parent quarks. The cross-section for the annihilation process can be obtained by adapting the cross-section given in Chapter 4 for $e^+ + e^- \rightarrow \mu^+ + \mu^-$. This yields

$$\sigma = (K/3) 4 Q_f^2 \pi \alpha^2 / 3M^2,$$

where the quark charge $Q_f e$ has replaced the electron charge e and the CM energy for the $q\bar{q}$ annihilation is the effective mass of the dilepton pair (M). An additional factor of 1/3 appears because the colours of the quark and antiquark from the two hadrons must match if they are to annihilate. Finally, K is the factor by which the cross-section is boosted when a leading order QCD calculation is made. K takes account of gluon radiation by the active quark and antiquark. K is predicted to be about 2·5 and fortunately it is sensibly constant for most of the kinematic region of interest.

If the Drell-Yan process does factorize as shown in Fig. 9.5 the Drell-Yan cross-section can be obtained by multiplying the above cross-section (i) by the probability $q_B(x_1)\, dx_1$ that a quark in the beam hadron carries a momentum fraction x_1, and (ii) by the probability $\bar{q}_T(x_2)\, dx_2$ that an antiquark in the target hadron carries a fraction x_2 of the target momentum. Then

$$d^2\sigma = (4\pi K \alpha^2 Q_f^2 / 9M^2) q_B(x_1) \bar{q}_T(x_2)\, dx_1\, dx_2.$$

Equally the quark may originate from the target and the antiquark from the beam, so there is a contribution with

$\bar{q}_B(x_1)q_T(x_2)$ replacing $q_B(x_1)\bar{q}_T(x_2)$. The full expression for the Drell-Yan differential cross-section is

$$d^2\sigma/dx_1dx_2=(4\pi K\alpha^2/9M^2)S(x_1, x_2),$$

where a sum over quark flavours (f) is made

$$S(x_1, x_2)=\Sigma Q_f^2[q_B(x_1)\bar{q}_T(x_2)+\bar{q}_B(x_1)q_T(x_2)].$$

The cross-section which has emerged shows some very simple properties. Re-expressing it in terms of the observables x_F and M,

$$d^2\sigma/dx_F\, dM=(8\pi K\alpha^2/9M^3)(\tau/\sqrt{x_F^2+4\tau})\, S(x_1, x_2).$$

Thus the quantity $M^3\, d\sigma/dM$ depends only on the structure of the parents. This scaling behaviour as well as the spectacular fall in $d\sigma/dM$ with M^3 are seen in the data from proton-proton collisions shown in Fig. 9.6 for $\sqrt{s}$ values from 23·8 to 62 GeV. The full line is a prediction made using the Drell-Yan model. Detailed comparisons have also been made between the measured Drell-Yan cross-sections for proton-proton collisions and predictions using structure functions measured in deep inelastic scattering experiments. Agreement is excellent and supports the concepts underlying the Drell-Yan calculations; namely the quark model, QCD and factorization. This then justifies extending the study to Drell-Yan production with π-meson beams. π-meson structure functions have been obtained from data on the reactions:

$$\pi^+(\pi^-)+\text{nucleus}\rightarrow\mu^++\mu^-+\text{anything}.$$

The most noteworthy feature observed is that the mean valence-quark momentum is about 1·5 times higher in mesons (which contain a valence quark and antiquark) than in nucleons (which contain three valence quarks). The fraction of momentum carried by gluons and the sea is similar to the nucleon case. The observed nucleon and meson valence-quark distributions (at low $|q^2|$) are crudely fitted by the expressions

$$q_v\ (\text{nucleon})\approx(1-x)^3/\sqrt{x}$$

$$q_v\ (\text{meson})\approx(1-x)/\sqrt{x}$$

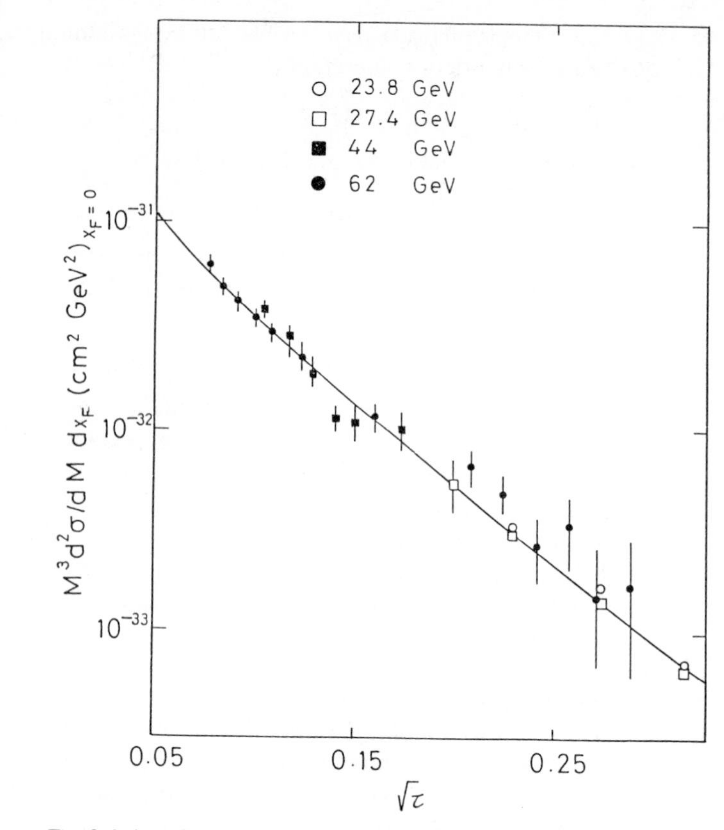

Fig. 9.6 A scaling test for proton-produced dimuons: measurements of the cross-section $M^3(d^3\sigma/dM\, dx_F)$ at $x_F=0$ are compared over a range of $\sqrt{s}$ from 23.8 to 62 GeV (after Kenyon, *Rep. Prog. Phys.* **45**, 1213 (1982)).

while for the sea:

$$q_s\text{ (nucleon)} \approx (1-x)^7/x$$
$$q_s\text{ (meson)} \approx (1-x)^5/x.$$

These parametric forms are discussed by Kenyon (1982).

A final point of interest is that the Drell-Yan process is responsible for the Z^0 and $W^{\pm}$ production seen in $\bar{p}p$ collisions.

In Fig. 9.5 the virtual photon would need to be replaced by a real Z^0 or $W^\pm$ in order to describe the reaction. Predictions for the W- and Z^0-production cross-sections (times the branching ratio for decay) have been made using standard structure functions. The predicted and measured cross-sections for proton-antiproton collisions at a CM energy of 540 GeV are compared in Table 9.1. They seem to agree well.

Table 9.1 *Production cross-sections for the W and Z^0 from $\bar{p}p$ annihilation at* 540 *GeV*

	Predicted cross-section x *branching fraction (pico-barns)*	*Observed cross-section* x *branching fraction (pico-barns)*
$W \rightarrow e\nu$	380^{+120}_{-50}	550 ± 100
$Z \rightarrow e^+e^-$	41^{+13}_{-7}	40 ± 20

Chapter 10
Hadronic flavour

The emphasis changes here from direct measurements of the internal structure of hadrons to the analysis of their spectra, quantum numbers and interactions. These properties also shed light on the internal structure of hadrons and serve to confirm the picture we have of the hadron composed of quarks and antiquarks held together by gluon exchanges. Quark flavour is a key quantum number because it is conserved in strong interactions. Each quark carries one of the six distinct flavours known at present (u, d, c, s, t, b) and it is evident that a bewildering array of hadrons of assorted flavours can be constructed from permutations of such quarks. The hadrons whose valence quarks (and antiquarks) are in s-states of relative motion have the lowest mass for a given flavour combination (e.g. π-meson and nucleons). When angular or radial modes of quark motion are excited the hadrons become more massive. Such excited states can generally decay via the strong interaction to lighter s-state hadrons. These transient excited states are also known as resonances. The static properties of hadrons are a principal theme of this chapter with something being said both about their measurement and interpretation.

The known hadron and lepton states are tabulated biennially and the latest version can be found in *Physics Letters* **170B** (1986). A limited selection of this data is presented in Appendix G.

10.1 Electron-positron annihilation to vector mesons

The annihilation of an electron and a positron each of energy E in a head-on collision gives a virtual photon of four-momentum

$(2E, 0)$. This can subsequently convert to a quark-antiquark pair and thence to hadrons (Fig. 6.1). In section 6.1 we saw how R, the hadron/lepton production cross-section ratio, increases by $3Q_f^2$ on crossing the threshold CM energy $(2m_f)$ for producing a quark-antiquark pair carrying charges $\pm Q_f e$ and mass m_f. At 2 GeV CM energy only the u-, d-, s-quark pairs can be produced and R is around 2. Fig. 10.1 shows that R exhibits a series of resonant peaks above 3 GeV and then settles at a new plateau value. The conclusion is that a threshold for a new flavour has been crossed, namely for producing charm quark-antiquark pairs. The mass of the charm quark is therefore about 1·5 GeV/c^2. If the charm quark's charge is $+2/3e$ this would lead to a rise of 4/3 in R. However, at 3·568 GeV the threshold has also been crossed for producing τ-lepton pairs. The τ-lepton decay products include hadrons 65 per cent of the time so that τ-production contributes a further increment of around 0·65 to the rise in R, unless excluded experimentally. Thus the total measured rise in R is more like 2·0. At around 10 GeV there is another set of narrow peaks accompanied by a rise in R of about 1/3. There is just a single threshold in this case: that for producing beauty quark-antiquark pairs with quark charge $-e/3$. Evidently the mass of the beauty quark is about 5 GeV/c^2. Beyond this threshold up to the highest currently accessible energy (46 GeV) R remains effectively constant. The absence of any further rise in R shows that the mass of the top quark must exceed 23 GeV/c^2.

The individual peaks seen in Fig. 10.1 indicate the existence of states which are produced from virtual photons and are made up of charmed quarks. These states therefore have the quantum numbers of the photon ($J^{PC} = 1^{--}$) and quark content $c\bar{c}$. The two lowest of these vector meson states, $\psi(3097)$ and $\psi'(3685)$, have widths of 69 and 215 keV/c^2. Such widths are unexpectedly small for strong decays; by contrast the more massive ψ-states have widths of 50 MeV/c^2 or so. Another threshold effect is at work here. The $\psi(3097)$ and $\psi'(3685)$ can only decay strongly to mesons made up of light quarks. Such decays take place via the three-gluon annihilation process drawn in Fig. 10.2(a). Annihilation to one gluon is ruled out by colour

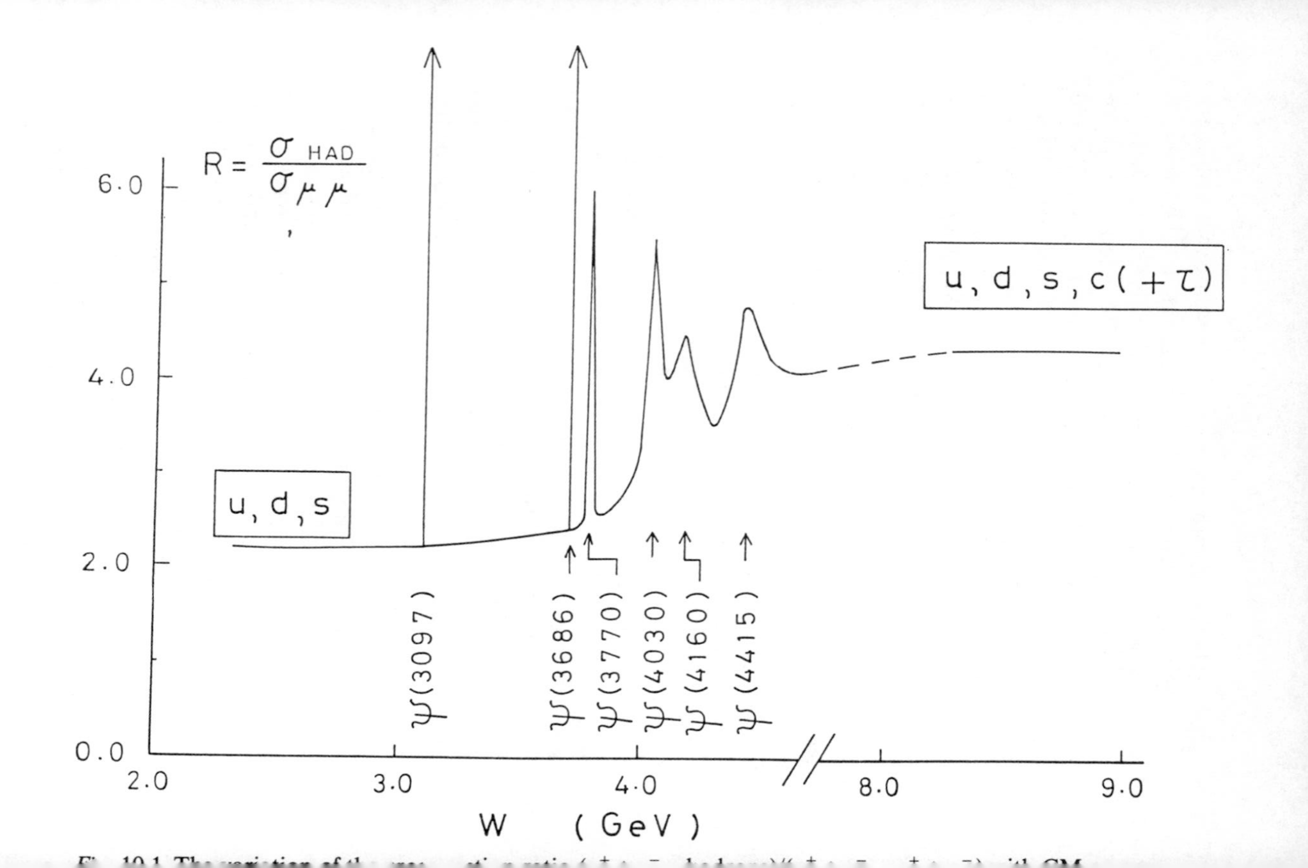
R = σ HAD / σ μ μ
u, d, s
u, d, s, c (+τ)
ψ(3097)
ψ(3686)
ψ(3770)
ψ(4030)
ψ(4160)
ψ(4415)
0.0
2.0
4.0
6.0
2.0
3.0
4.0
8.0
9.0
W (GeV)

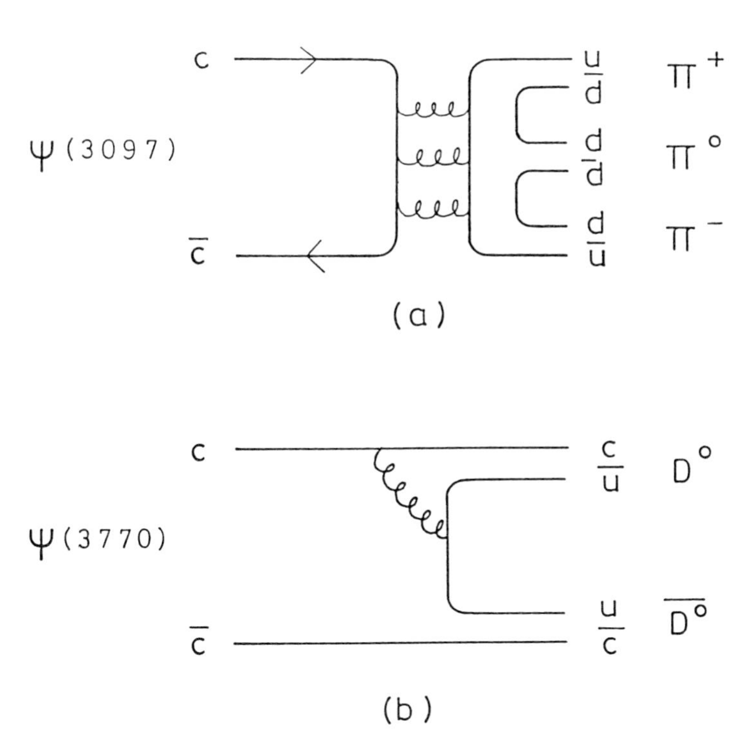

Fig. 10.2 (a) ψ(3097)-decay via annihilation to three gluons. (b) ψ(3770)-decay to D^0(1865) $\bar{D}^0$(1865).

conservation and annihilation to two gluons is forbidden by C parity conservation. The three-gluon decay has a rate proportional to $\alpha_s^6(M^2)$, where α_s is the QCD analogue of the fine-structure constant, introduced in Chapter 6. This quantity is very small at 3 GeV, and accounts for the suppression of the three-gluon decay. Above a mass of 3730 MeV/c^2 it is possible for the vector mesons to decay strongly to a D^0(1865)$\bar{D}^0$(1865) pair, as shown in Fig. 10.2(b), the D^0 and its antiparticle $\bar{D}^0$ being the lightest mesons containing a charmed quark or antiquark. The only gluon involved is now soft (low four-momentum) so that α_s is large and the decay is uninhibited. The suppression of processes in which there are only hard gluon links to one or more final-state hadrons is known as Zweig Rule suppression.

Mesons made up of $c\bar{c}$ or $b\bar{b}$ quarks have properties analogous to the bound e^+e^- system (positronium) and are called quarkonia. Quarkonium states with the quark and antiquark spins aligned parallel are called orthoquarkonia and those with the spins antiparallel are called paraquarkonia. The direct product of e^+e^- annihilations are states with $J^{PC}=1^{--}$, which are therefore orthoquarkonia. Fig. 10.3 shows all the observed orthoquarkonia (parallel spin) levels for $c\bar{c}$ and $b\bar{b}$ and their electromagnetic decays. The level spacings are small compared to the total mass, so we can infer that the quarks are moving non-relativistically. Simple non-relativistic potential model calculations of the level structure have been made based on potentials like that given in section 6.4 which incorporate the correct asymptotic behaviour at small and large distances. If the potential is flavour independent then the *s*-state levels of charm- and beauty-onia should have similar spacing, which is confirmed by the experimental data shown in Fig. 10.3. Using a potential model Buchmuller and Tye (1981) predicted the masses of quarkonia levels and these are indicated by the arrows on Fig. 10.3. There is overall very good agreement between the predicted masses and the observed masses for those states so far observed.

10.2 Flavour symmetries

Flavour symmetries arise because the current quark masses are light on the scale set by QCD (Λ). The current mass of a quark is the mass of a quark free of the dynamical effects felt in hadrons. It is the mass appearing in the Dirac equation for the quark and in QCD amplitudes. Gasser and Leutwyler (1982) present the following expression for the hadron mass M in terms of the current quark masses m_{f}

$$M^2 = M_0^2 + \Sigma A_{\mathrm{f}} m_{\mathrm{f}},$$

where M_0 would be the mass of the hadron if the quarks were massless. M_0 is smallest for the pseudoscalar mesons. Taking

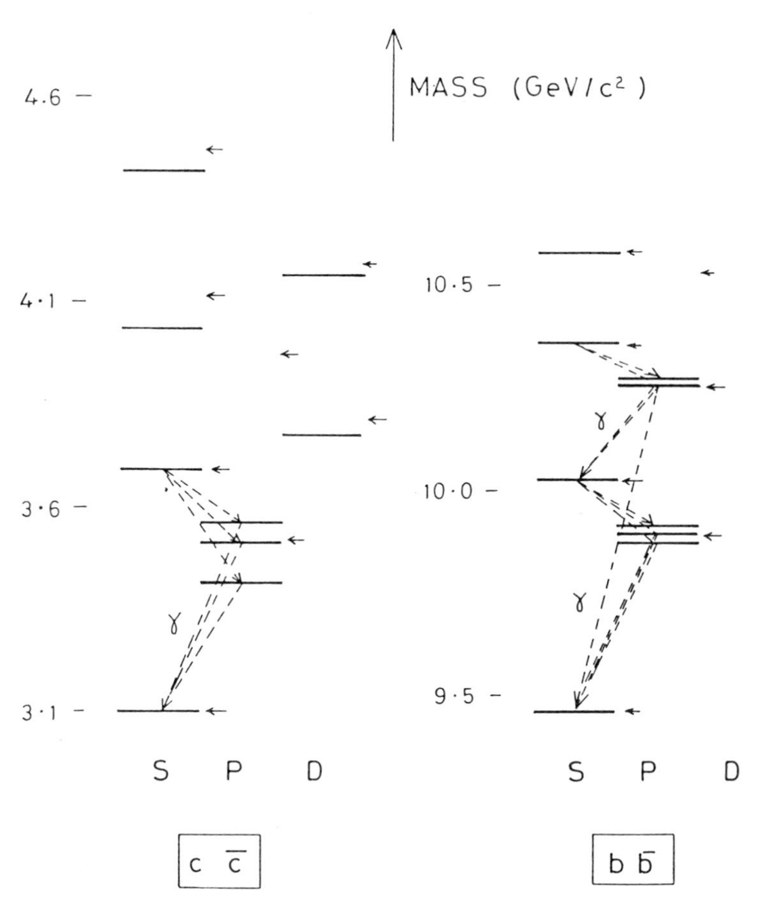

Fig. 10.3 Orthoquarkonia level structure for charm and beauty. The predictions of Buchmuller and Tye (1981) are indicated by horizontal arrows. Dotted lines show the observed electromagnetic decays.

the mass ratios we obtain

$$(m_s - \bar{m})/(m_d - m_u) = [M(K)^2 - M(\pi)^2]/[M(K^0)^2 - M(K^+)^2] = 57,$$

and if we ignore M_0

$$(m_s + \bar{m})/2\bar{m} = M(K)^2/M(\pi)^2 = 13,$$

where $\bar{m}$ is $(m_u + m_d)/2$. These relations yield $m_s = 25\bar{m}$ and $(m_d - m_u) = 0{\cdot}42\bar{m}$. According to the same authors the analysis of the neutron-proton mass difference yields a value of 3 MeV/c^2 for $(m_d - m_u)$. Table 10.1 summarizes the values deduced for the current masses. The *dynamical* mass of the u- and d-quarks is on the other hand obtained simply by dividing the nucleon mass by three. Then the regular increase of decuplet masses seen in Fig. 2.3(b) shows that s-quark is heavier by 140 MeV/c^2 than the u- or d-quark. Masses for the heavier quarks have been determined by fitting the quarkonia spectra; and from the observed masses of the charmed pseudoscalar mesons D^0(1865) and D^+(1869), and the beauty pseudoscalar mesons B^0(5275) and B^+(5271). These dynamical masses are also collected in Table 10.1.

Table 10.1

Quark species	*Current mass MeV/c²*	*Dynamical mass MeV/c²*
u	5·5	330
d	8·5	
s	150	470
c	1 250	1 580
b	4 250	4 580
t	—	(>40 000?)

10.2.1 Flavour SU(n) multiplets

The u- and d-quarks have current masses which are negligible compared to the energy scale ($\Lambda \sim 0{\cdot}2$ GeV) of the strong interaction and are therefore indistinguishable to the strong interaction. This feature is the origin of the strong-isospin symmetry SU(2), discussed in Chapters 2 and 5. This makes strong isospin an accidental symmetry, and indeed it does not seem to be connected with any new fundamental force. Why there should be a sequence of quark flavours with very different masses remains a mystery on which the standard model is silent. All the hadrons made from u- or d-quarks appear in strong-isospin multiplets with similar masses but differing charges. Strong isospin is conserved by the strong interaction but not by the electroweak interactions. It is possible to extend the flavour symmetry group to SU(3) by including s- as well as u- and d-quarks (Gell-Mann (1964), Zweig (1964)). However, the mass of the s-quark is comparable to Λ in magnitude so that the symmetry only holds approximately. The result is that multiplets of hadrons with identical quantum numbers (J^{PC}) are observed corresponding to the representations of flavour SU(3). Members of the same SU(3) representation can have very different masses and this in turn affects their interactions through the kinematics. The mathematical development for flavour SU(3) is identical to that for colour SU(3). There are two inequivalent fundamental representations, 3 and $\overline{3}$. Of these the first contains the quarks and the second the antiquarks. Figure 10.4 shows the respective weight diagrams using the flavour coordinates. Here

$$I_3 = F_3 \text{ (third component of strong isospin)}$$
$$Y = 2F_8/\sqrt{3} \text{ (strong hypercharge).}$$

Hypercharge is so named because ($Y/2$) is the mean charge carried by an isospin multiplet. The connection between charge and strong isospin has been given in Chapter 2:

$$Q = I_3 + Y/2 = I_3 + (B+S)/2,$$

where S is the quantum number known as strangeness, and B is

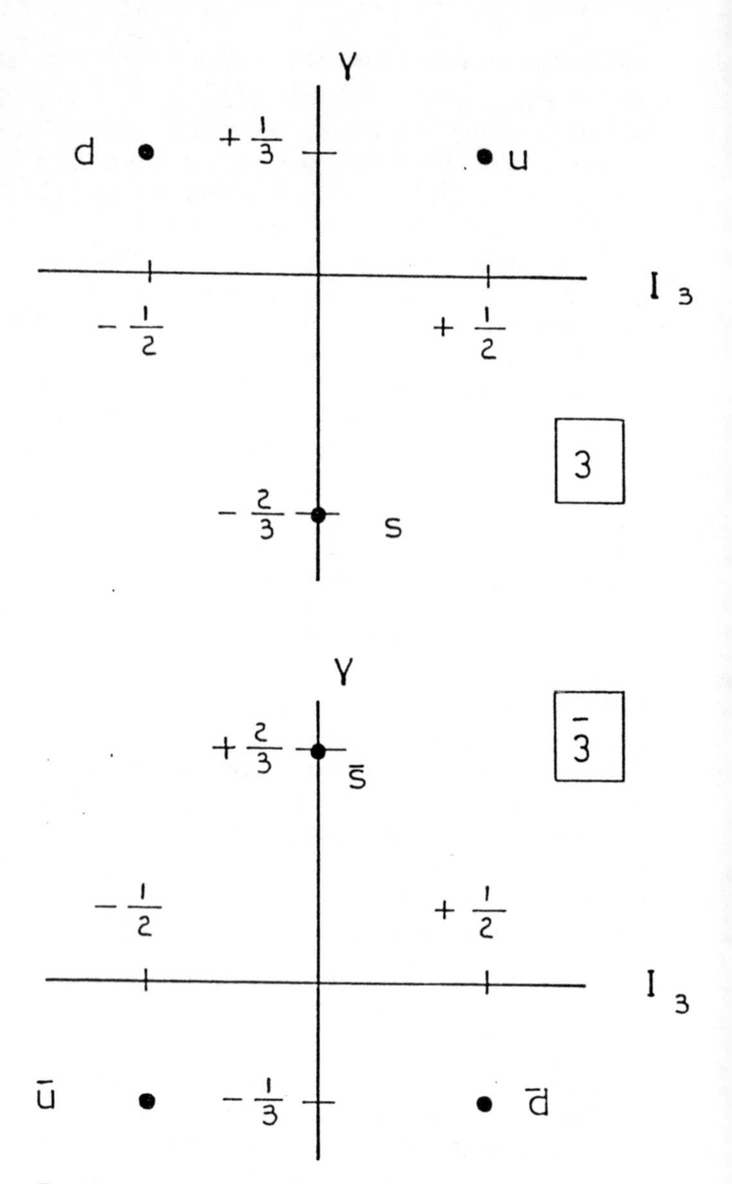

Fig. 10.4 The weight diagrams for the two inequivalent fundamental representations of flavour SU(3).

the baryon number (0 for mesons, +1 for baryons and, −1 for antibaryons). This relation parallels the expression found in Chapter 8 connecting charge with weak isospin and weak hypercharge. Table 10.2 (duplicate of Table 2.1) summarizes the quantum numbers for the u-, d- and s-quarks.

The stepping operators of SU(3) are these:

$$I_{\pm}=F_1 \pm iF_2; \quad V_{\pm}=F_4 \pm iF_5; \quad U_{\pm}=F_6 \pm iF_7.$$

With the usual notation:

$$u=\begin{bmatrix}1\\0\\0\end{bmatrix}, d=\begin{bmatrix}0\\1\\0\end{bmatrix} \text{ and } s=\begin{bmatrix}0\\0\\1\end{bmatrix},$$

so that $I_+d=u$, $I_-u=d$, $V_+s=u$, $V_-u=s$, $U_+s=d$, $U_-d=s$, while all other operations annihilate the quarks.

The mesons are constructed from quark-antiquark pairs and belong to the irreducible representations formed from decomposing the product $3 \otimes \overline{3}$, namely from $1 \oplus 8$. Figure 10.5 shows the weight diagram for the octet and singlet states. These states are obtained by taking the $\overline{3}$ weight diagram and centering it in turn on each of the weights in the 3 weight diagram. The lowest mass states have the quark-antiquark pair in an *s*-state with either antiparallel or parallel spins. The quark model predicts accurately that these states have J^{PC} of 0^{-+} and 1^{--} respectively (section 5.7). The nine observed pseudoscalar and vector mesons are named on Fig. 10.5. We note the wide dispersion in mass coming from the mass difference between u- and d-quarks on the one hand and s-quarks on the other. The

Table 10.2

quark	Q	I	I_3	Y	S	B
u	$+\frac{2}{3}$	$\frac{1}{2}$	$+\frac{1}{2}$	$\frac{1}{3}$	0	$+\frac{1}{3}$
d	$-\frac{1}{3}$	$\frac{1}{2}$	$-\frac{1}{2}$	$\frac{1}{3}$	0	$+\frac{1}{3}$
s	$-\frac{1}{3}$	0	0	$-\frac{2}{3}$	−1	$+\frac{1}{3}$

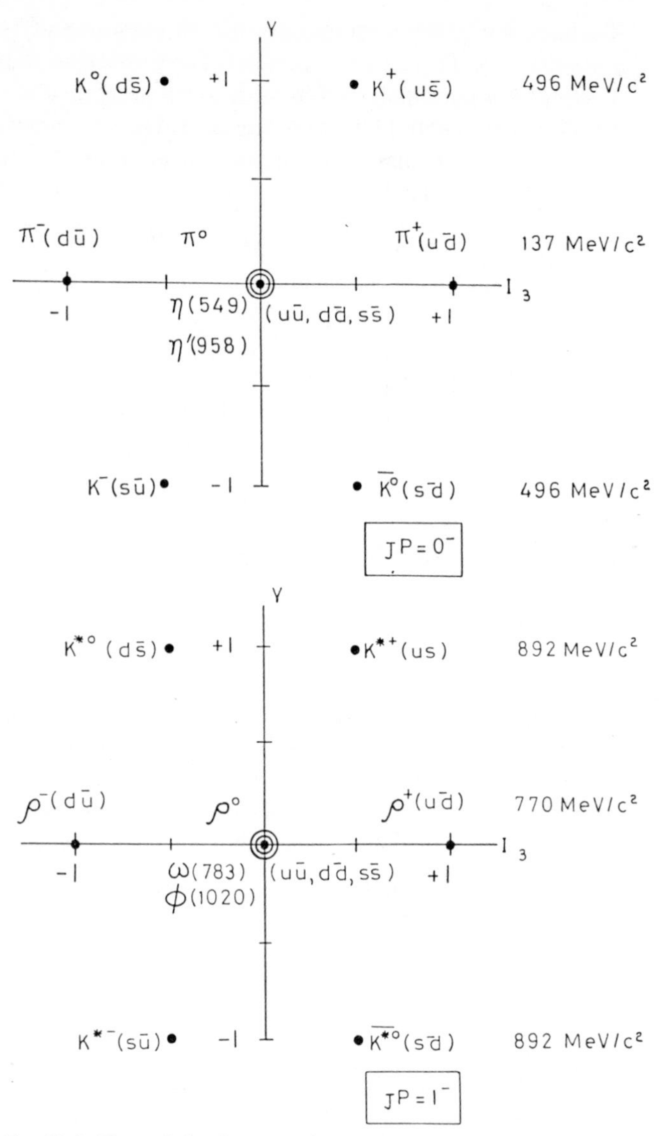

Fig. 10.5 The weight diagrams for the octet and singlet states formed from a quark plus an antiquark, (a) pseudoscalar mesons, (b) vector mesons. The mean masses of each multiplet are also shown.

pseudoscalars range in mass from the $\pi^0(135)$ to the $\eta'(958)$, where the arguments are the particle masses in MeV/c^2. Four further sets of octets plus singlets are observed experimentally in the mass region 950–1550 MeV/c^2. These are states with unit relative orbital angular momentum between quark and antiquark. They are the ${}^1P(J^{PC}=1^{+-})$ and ${}^3P(J^{PC}=0^{++}, 1^{++}, 2^{++})$ nonets. At higher masses the increasing density of states and the falling production cross-sections hamper investigation. Nonetheless several d- and f-wave states have been observed. A few examples of hadrons with radial excitation are also known; e.g. the $\pi(1300)$. Not only does the meson spectrum contain the spin-parity combinations expected for quark-antiquark structures but also those forbidden are notably absent. Thus, no examples of doubly charged mesons have been detected, nor cases of the J^{PC} combinations $(\text{odd})^{-+}$, $(\text{even})^{+-}$ nor 0^{--}.

The baryon spectrum also follows the quark-model predictions. Decomposing the product of flavour representations for three quarks gives

$$3\otimes 3\otimes 3 = 1\oplus 8\oplus 8\oplus 10.$$

In their ground state the quarks will have zero relative orbital angular momentum. Possible spin combinations are thus restricted to $\frac{1}{2}$ or $\frac{3}{2}$ and positive parity is required for these ground states. The observed lowest mass baryons are an octet with spin-parity $\frac{1}{2}^+$ and a decuplet with spin-parity $\frac{3}{2}^+$. Both multiplets have been shown earlier in Fig. 2.3 on SU(3) weight diagrams.

An extension to include four-quark species (u, d, s, c) gives an SU(4) symmetry which is very badly broken because $m_c \gg \Lambda$. The lower lying multiplets of mesons are obtained from the product

$$4\otimes\bar{4} = 15\oplus 1.$$

All the members of the 0^- 15-plet have been observed. They are:

3 π-mesons: $\pi^+(140)\equiv \mathrm{u\bar{d}}$, $\quad \pi^0(135)\equiv(\mathrm{d\bar{d}}-\mathrm{u\bar{u}})/\sqrt{2}$

$\pi^-(140)\equiv \mathrm{\bar{u}d}$

4 K-mesons: $K^+(494) \equiv u\bar{s}$, $K^0(498) \equiv d\bar{s}$

$\bar{K}^0(498) \equiv \bar{d}s$, $K^-(494) \equiv \bar{u}s$

1 η_8-meson: $\eta_8 \equiv (u\bar{u} + d\bar{d} - 2s\bar{s})/\sqrt{6}$

4 D-mesons: $D^+(1869) \equiv c\bar{d}$, $D^0(1865) \equiv c\bar{u}$

$\bar{D}^0(1865) \equiv \bar{c}u$, $D^-(1869) \equiv \bar{c}d$

2 D_s-mesons: $D_s^+(1971) \equiv c\bar{s}$, $D_s^-(1971) \equiv \bar{c}s$

1η_c-meson: $\eta_c(2981) \equiv c\bar{c}$.

With five species of quark (u, d, s, c, b) there is a vestigial SU(5) flavour symmetry having a pseudoscalar 24-plet and a singlet. The only examples of 0^- mesons so far seen containing a b-quark are

4 B-mesons: $B^+(5271) \equiv \bar{b}u$, $B^0(5275) \equiv \bar{b}d$

$\bar{B}^0(5275) \equiv b\bar{d}$, $B^-(5271) \equiv b\bar{u}$.

Five other pseudoscalar possibilities ($b\bar{s}$, $\bar{b}s$, $c\bar{b}$, $\bar{c}b$, $b\bar{b}$) have yet to be observed.

10.2.2 Light hadron masses

Light hadrons do not lend themselves to the potential model calculations used for quarkonia because the quark-antiquark motion is relativistic and the quark-antiquark separations are larger. Qualitative effects due to quark mass differences, electromagnetic effects, and QCD fine and hyperfine splitting are all seen. Two examples are now discussed in detail.

The first example is the level splitting between the *s*-wave baryon masses, due to two effects: the differences between quark masses, and the hyperfine (spin-spin) splitting. In the absence of hyperfine splitting the $\Delta(1232)$ mass and nucleon mass would converge to 1085 MeV/c^2. Successive replacements of the u- or d-quarks by s-quarks would increase this mass by 140 MeV/c^2 steps per unit change in strangeness, as shown in the centre column of Fig. 10.6. De Rujula *et al.* (1975) have developed the idea that one-gluon exchange would introduce an effective interaction with the same formal structure as that observed for

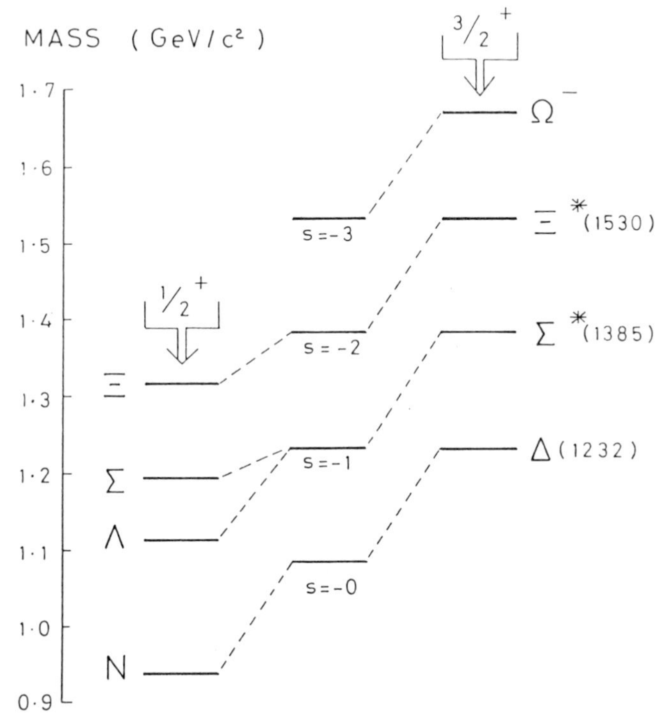

Fig. 10.6 The mass spectrum of the *s*-wave baryons. In the centre column are shown the levels before hyperfine splitting.

one-photon exchange in atoms or in positronium. Of the energy terms expected those due to the spin-orbit and tensor spin-spin forces vanish in the *s*-wave state of relative motion leaving only a contribution from the hyperfine interaction between the magnetic moments of the quarks inside the hadron. This contribution to the energy is:

$$E_{\mathrm{HF}} = -\sum_{ij} \langle(\mathbf{s}_i \cdot \mathbf{s}_j)\rangle \langle(\mathbf{F}_i \cdot \mathbf{F}_j)\rangle / m_i m_j$$

where i and j are the quark labels and where $\mathbf{s}$, $\mathbf{F}$ and m are the quark spin, colour generator and mass respectively. The sum is over all quark (antiquark) pairs. For comparison in the atomic case the hyperfine and fine structure energy terms are

$$E_{\mathrm{HF}} \sim e^2 \mathbf{s}_i \cdot \mathbf{s}_j / m_{\mathrm{e}} m_{\mathrm{p}}, \quad E_{\mathrm{F}} \sim e^2 \mathbf{L} \cdot \mathbf{s} / m_{\mathrm{e}}^2.$$

Notice that the appropriate expectation value of the colour generator has replaced the charge in the transition from QED to QCD. In an atom the hyperfine splitting is suppressed compared to fine structure because m_{p} is 2 000 times larger than m_{e}. However, the hyperfine splitting is not suppressed in the case of gluon exchange because all the masses appearing are quark masses. Expanding the colour term,

$$2\mathbf{F}_i \cdot \mathbf{F}_j = (F_{\mathrm{i}} + F_{\mathrm{j}})^2 - F_i^2 - F_j^2,$$

where F_i^2 and F_j^2 are to be evaluated for the quark 3 representation. Each pair of quarks must form a colour $\overline{3}$ in order that when combined with a third quark a colour singlet should result. Recall that $3 \otimes 3 = \overline{3} \oplus 6$. This means that $F^2 = (F_{\mathrm{i}} + F_{\mathrm{j}})^2$ must be evaluated for the $\overline{3}$ representation. Then, using the values given for $\langle F^2 \rangle$ in Table 6.1,

$$\langle 2\mathbf{F}_i \cdot \mathbf{F}_j \rangle = -4/3.$$

A similar analysis gives

$$2\sum_{ij} \langle \mathbf{s}_i \cdot \mathbf{s}_j \rangle = \langle (\mathbf{s}_1 + \mathbf{s}_2 + \mathbf{s}_3)^2 - s_1^2 - s_2^2 - s_3^2 \rangle.$$

The quark spins s_1, s_2 and s_3 are all $\frac{1}{2}$ so that the expectation values of s_1^2 etc., are all $\frac{1}{2}(\frac{1}{2}+1)$, i.e. 3/4. Thus

$$2\sum_{ij} \langle \mathbf{s}_i \cdot \mathbf{s}_j \rangle = s(s+1) - 3(3/4),$$

where s is the baryon spin. Therefore

$$2\Sigma \langle \mathbf{s}_i \cdot \mathbf{s}_j \rangle = +3/2 \text{ for the decuplet, } -3/2 \text{ for the octet.}$$

E_{HF} is overall positive for the decuplet and negative for the octet. We see that the hyperfine interaction displaces the decuplet upward and the octet downward through equal mass intervals. There is some reduction in this effect with increasing strangeness because the light u- and d-quarks are being replaced by the heavier s-quarks. The final observed levels are shown in the right- and left-handed columns in Fig. 10.6. The same form for the hyperfine interaction also leads to a consistent explanation of the mass splitting between the 0^- and 1^- octets of mesons.

The second example of a qualitative mass effect to be considered here concerns the vector mesons and reveals that there is mixing between the octet and singlet $I_3 = Y = 0$ states. To begin with we write the quark structure of the neutral members of the octet:

$$\rho^0 \equiv (u\bar{u} - d\bar{d})/\sqrt{2}, \quad \mathrm{K}^{*0} \equiv d\bar{s},$$

and the $I=0$ member $\omega_8 \equiv (u\bar{u} + d\bar{d} - 2s\bar{s})/\sqrt{6}$.
These have masses given by the expression in section 10.2:

$$\begin{aligned} m^2(\rho) &= m_\mathrm{u} + m_\mathrm{d} + M_0^2, \\ m^2(\mathrm{K}^*) &= m_\mathrm{d} + m_\mathrm{s} + M_0^2 \quad \text{and} \\ m^2(\omega_8) &= (m_\mathrm{u} + m_\mathrm{d} + 4m_\mathrm{s})/3 + M_0^2, \end{aligned}$$

where m_u, m_s and m_d are the current quark masses. Then ignoring the small difference between m_u and m_d

$$m^2(\rho) + 3m^2(\omega_8) = 4m^2(\mathrm{K}^*).$$

Inserting the experimentally determined masses of the $\rho(770)$ and $\mathrm{K}^*(890)$ in this equation yields

$$m(\omega_8) = 930 \text{ MeV}/c^2,$$

which lies midway in mass between the observed $I_3 = Y = 0$ states, $\omega(780)$ and $\phi(1020)$. The most economic explanation of this discrepancy is to say that the singlet ω_1 and octet ω_8 mix to form the physical $\omega(780)$ and $\phi(1020)$:

$$\begin{aligned} |\omega\rangle &= |\omega_1\rangle \cos\theta_\mathrm{v} + |\omega_8\rangle \sin\theta_\mathrm{v} \\ |\phi\rangle &= |\omega_1\rangle \sin\theta_\mathrm{v} - |\omega_8\rangle \cos\theta_\mathrm{v} \end{aligned}$$

In terms of the physical states the mass matrix is diagonal, so

$$\begin{aligned} m^2(\omega) &= \langle\omega|m^2|\omega\rangle \\ &= m^2(\omega_1)\cos^2\theta_\mathrm{v} + m^2(\omega_8)\sin^2\theta_\mathrm{v} \\ &\quad + 2\langle\omega_1|m^2|\omega_8\rangle \cos\theta_\mathrm{v} \sin\theta_\mathrm{v} \end{aligned}$$

and

$$m^2(\phi)=m^2(\omega_1)\sin^2\theta_v+m^2(\omega_8)\cos^2\theta_v$$
$$-2\langle\omega_1|m^2|\omega_8\rangle\cos\theta_v\sin\theta_v.$$

Also

$$\langle\omega|m^2|\phi\rangle=0$$

i.e.

$$m^2(\omega_1)\cos\theta_v\sin\theta_v-m^2(\omega_8)\cos\theta_v\sin\theta_v$$
$$+\langle\omega_1|m^2|\omega_8\rangle(\sin^2\theta_v-\cos^2\theta_v)=0.$$

Eliminating $\langle\omega_1|m^2|\omega_8\rangle$ and $m^2(\omega_1)$ yields

$$\tan^2\theta_v=\frac{m^2(\omega_8)-m^2(\phi)}{m^2(\omega)-m^2(\omega_8)}.$$

When the masses of ω_8, ω and ϕ are substituted into this expression, θ_v is found to be about 39°. This is close to the 'ideal' mixing angle ($\tan\theta_v=1/\sqrt{2}$ and $\theta_v=35{\cdot}3$) at which the mesons are made either from only the lighter u- and d-quarks or from the heavier s-quarks only:

$$\omega=(uu+dd)/\sqrt{2},\quad \phi=ss.$$

One consequence of ideal mixing is that the strong decay of the ϕ to π-mesons is suppressed by the Zweig Rule (compare Fig. 10.2). Decay to K-mesons is permitted by this rule but is restrained by the small energy release (~25 MeV). The ϕ-meson therefore has the rather narrow width of 4·2 MeV/c^2 for what is a strong decay.

10.3 The determination of meson quantum numbers

Information on the excited hadron states and their quantum numbers underpins the quark model. Some simple cases of their determination will be discussed in this and the following sections. A typical hadron excited state (e.g. $\Delta(1232)$) has a width (Γ) 0·1 GeV/c^2 and hence a lifetime of $6{\cdot}58\times10^{-25}/0{\cdot}1\approx10^{-23}$ s. Its transient existence precludes direct detection. If the excited hadron state with four-momentum $(E_x, \mathbf{P}_x)$ decays to daughters labelled, $i=1, \ldots n$ then its mass is given by

$$M_x^2 = E_x^2 - \mathbf{P}_x \cdot \mathbf{P}_x = \sum_i (E_i)^2 - \left(\sum_i \mathbf{P}_i\right)^2.$$

The quantity on the right-hand side is called the *effective mass squared* of the daughters. It follows that an excited hadron will show itself as a Breit-Wigner resonance (Appendix B) in the effective mass of its daughters. A simple example is provided by the reaction:

$$\pi^- + \mathrm{p} \rightarrow \pi^+ + \pi^- + \mathrm{n}$$

when the π-meson beam energy is tens of GeV. The effective mass spectrum of the outgoing pair of π-mesons (Fig. 10.7) exhibits a series of resonant peaks each corresponding to the decay of an excited meson; $\rho(770)$, $\mathrm{f}_2(1260)$, $\mathrm{g}_3(1690)$ and $\mathrm{f}_4(2030)$. Events for which the effective mass of the two π-mesons is close to 770 MeV/c^2 involve a two-step process of ρ-production and decay

$$\pi^- + \mathrm{p} \rightarrow \rho^0 + \mathrm{n}$$
$$\hookrightarrow \pi^+ + \pi^-.$$

Figure 10.7 also shows the effective mass spectrum for $\pi^-\pi^0$ pairs produced in the reaction

$$\pi^- + \mathrm{p} \rightarrow \pi^- + \pi^0 + \mathrm{p}.$$

In this case only the resonant peaks corresponding to ρ^-- and g_3^--production are present. The resonance production in all the related channels is summarized below:

$\pi^- + \mathrm{p} \rightarrow \pi^+ + \pi^- + \mathrm{n}$	$\rho \mathrm{f}_2 \mathrm{g}_3 \mathrm{f}_4$
$\pi^\pm + \mathrm{p} \rightarrow \pi^\pm + \pi^0 + \mathrm{p}$	$\rho - \mathrm{g}_3 -$
$\pi^- + \mathrm{p} \rightarrow \pi^0 + \pi^0 + \mathrm{n}$	$- \mathrm{f}_2 - \mathrm{f}_4$
$\pi^+ + \mathrm{p} \rightarrow \pi^+ + \pi^+ + \mathrm{n}$	$- - - -$

The f_2 and f_4 are observed only in neutral combinations ($\pi^+\pi^-$ or $\pi^0\pi^0$) and thus have isospin zero. ρ and g_3 are observed in

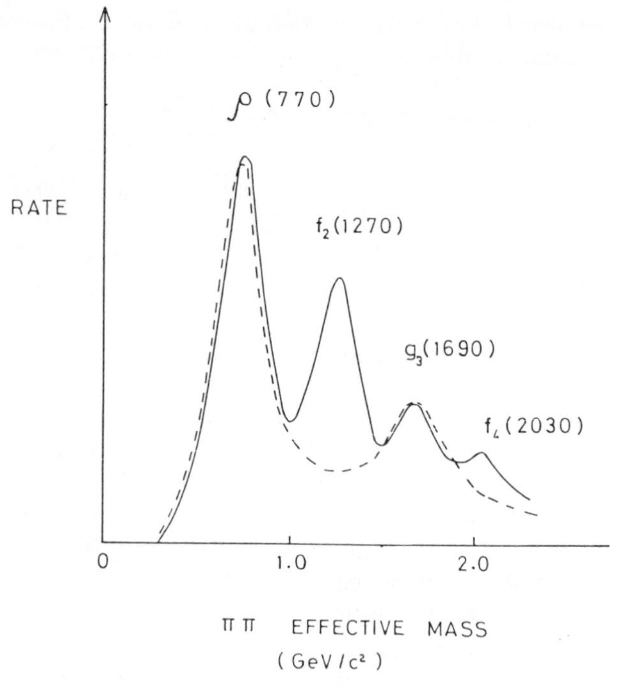

Fig. 10.7 A sketch of the $\pi-\pi$ effective mass spectra observed in the reactions $\pi^- + p \rightarrow \pi^+ + \pi^- + n$ (full line) and $\pi^- + \pi^0 + p$ (broken line).

singly charged and neutral combinations ($\pi^+\pi^-$ or $\pi^0\pi^0$) but not in $\pi^+\pi^+$; they have unit strong isospin. The decay of ρ^0 or g_3^0 to $\pi^0\pi^0$ is not possible because the Clebsch-Gordan coefficient $(1, 0|1, 0; 1, 0)$ for the isospin overlap is identically zero. Finally, the absence of any peaks in the $\pi^+\pi^+$ effective mass spectrum indicates that there are no meson excited states with strong isospin equal to 2 or greater.

The arguments used in section 5.7 to obtain the quark-antiquark parity and *C*-parity, can be applied to a neutral π-meson pair with relative orbital angular momentum L. We make use of the fact that anti π-mesons are also π-mesons; anti (π^-) is π^+ and anti(π^0) is π^0. A meson-antimeson pair has positive intrinsic parity. In addition the π-mesons are spinless so that

$$C = P = (-1)^J \quad \text{and} \quad J = L.$$

If anti π-mesons are also π-mesons then the overall wavefunction $\phi(\text{space}) \times \phi(\text{isospin})$ must be symmetric under meson-antimeson interchange because of Bose symmetrization. Thus for even orbital angular momentum the strong-isospin state must be symmetric, i.e. I must be zero. Similarly odd orbital angular momentum requires the strong isospin to be unity. Collecting these results:

Charge combination	*Allowed $I(J^{PC})$*
$\pi^+\pi^-$	$\begin{cases} 0(0^{++}, 2^{++}, \ldots) \\ 1(1^{--}, 3^{--}, \ldots) \end{cases}$
$\pi^0\pi^0$	$0(0^{++}, 2^{++}, \ldots)$

The ρ- and g_3-mesons must have J^{PC} in the sequence $1^{--}, 3^{--}, \ldots$ and unit strong isospin; the f_2- and f_4-mesons must have J^{PC} in the sequence $0^{++}, 2^{++}, \ldots$ and zero strong isospin. In order to distinguish between the alternative choices of J^{PC} it is necessary to measure the angular distribution of the π-mesons emitted in the parent rest frame. Suppose the ρ-mesons are produced in a pure state $|JM\rangle$, where J is the ρ-meson spin and M its component along a quantization axis. The expansion of such a state in terms of two spinless particles (the daughter π-mesons) is given in section 5.5:

$$|JM\rangle = \int \mathrm{d}\Omega \, Y_J^M(\alpha, \beta) |\alpha\beta\rangle,$$

where $|\alpha\beta\rangle$ are plane-wave states with the π-mesons travelling back-to-back along directions with polar angles (α, β) and $(\pi - \alpha, \beta + \pi)$ with respect to the quantization axis. Hence the amplitude for emission along the direction (θ, ϕ) is

$$\langle\theta\phi|JM\rangle = \int \mathrm{d}\Omega \, Y_J^M(\alpha, \beta) \langle\theta, \phi|\alpha, \beta\rangle = Y_J^M(\theta, \phi)$$

and the intensity is

$$I(\theta, \phi) = |Y_J^M(\theta, \phi)|^2.$$

More generally the ρ-meson can be produced in a state which is a linear superposition of magnetic substates each with coefficient a_{M}. Then the intensity

$$\begin{aligned} I(\theta, \phi) &= |\Sigma a_{\mathrm{M}} Y_J^M(\theta, \phi)|^2 \\ &= \Sigma b_{\mathrm{LN}} Y_L^N(\theta, \phi), \end{aligned}$$

where $0 \leqslant L \leqslant 2J$ and $-L \leqslant N \leqslant L$ with coefficients b_{LN} which are bilinear combinations of the coefficients a_{M}. The coefficients b_{LN} can easily be obtained from the measured angular distribution $I(\theta, \phi)$ by taking advantage of the orthogonal property of the spherical harmonic functions. The integral is formed from the data

$$\int I(\theta, \phi) Y_L^{N*}(\theta, \phi) \mathrm{d}\Omega = N b_{\mathrm{LN}},$$

where N is the number of events in the data sample. Next the a_{N} coefficients of interest can be extracted from these b_{LN} coefficients. A practical complication is that the meson which decays may be produced in an incoherent mixture of magnetic substates; for this situation it is necessary to use the density matrix formalism (see Williams (1971)). Analyses of the type just described show that the ρ, f_2, g_3 and f_4 have J^{PC} equal to 1^{--}, 2^{++}, 3^{--} and 4^{++} respectively. They form a sequence of $\mathrm{q\bar{q}}$ states with increasing angular momentum. The analysis of many-body decays exploits quite similar techniques.

10.4 The determination of baryon quantum numbers

Baryon quantum numbers have often been determined in what are called *formation* experiments, in which a meson beam bombards a hydrogen or deuterium target and the beam meson plus target nucleon form a resonant state. The partial-wave analysis of this process will be described here in terms of scattering from a potential. We start by considering the case of elastic scattering of spinless particles.

A plane wave is incident travelling along the z-direction toward the scattering centre at 0. At distances remote from 0 and in the absence of the scatterer the following expansion is valid

$$\psi_o \equiv \exp(i(kz-\omega t)) = (i/2kr)\sum_L (2L+1)\{(-1)^L \exp(-ikr) - \exp(ikr)\}P_L^0(\cos\theta)\exp(-i\omega t).$$

Here k is the momentum and ω the energy of a beam particle and r is the distance from the scatterer. The right-hand side of the equation consists of an incoming spherical wave $\exp(-i(kr+\omega t))$ and an outgoing spherical wave $\exp(i(kr-\omega t))$. The incident plane wave has no component of orbital angular momentum (L_z) along its direction of motion $0z$. This statement is easy to prove:

$$L_z\psi_0 \equiv -i(x\partial/\partial y - y\partial/\partial x)\psi_0 = 0$$

because ψ_0 depends only on z. Correspondingly on the RHS the Legendre polynomials (P_L^0) are those with the z-component of orbital angular momentum, M, equal to zero. Only the outgoing waves are at all affected by the scatter. For the L'th *partial wave* (i.e. the wave with orbital angular momentum L) the result is in general to reduce the outgoing wave by a factor $\eta_L (0 \leqslant \eta_L \leqslant 1)$ and delay its phase by $2\delta_L$. Then the wave becomes

$$\psi_1 = (i/2kr)\sum_L (2L+1)\{(-1)^L \exp(-ikr) - \eta_L \exp i(kr+2\delta_L)\}P_L^0(\cos\theta)\exp(-i\omega t).$$

The elastically scattered wave is the difference between ψ_1 and ψ_0, i.e.

$$\psi_{sc} = (1/2ik)\,\Sigma(2L+1)\{\eta_L e^{2i\delta_L} - 1\}\,(\exp(ikr)/r)P_L^0(\cos\theta)$$
$$= f(\theta)\exp(ikr)/r,$$

where the time dependence has been removed. The flux of elasticity scattered into a solid angle $d\Omega$ is

$$|\psi_{sc}|^2 vr^2 \, d\Omega = |f(\theta)|^2 v \, d\Omega,$$

where v is the velocity of both incident and scattered particles.

The differential cross-section is this flux divided by the beam flux v; i.e.

$$d\sigma = |f(\theta)|^2 \, d\Omega.$$

$$\therefore d\sigma/d\Omega = (1/k^2)|\Sigma(2L+1)T_L P_L^0(\cos\theta)|^2.$$

The coefficient

$$T_L = (\eta_L \exp(2i\delta_L) - 1)/2i$$

is called the *elastic partial-wave amplitude*. (For historical reasons this partial-wave amplitude differs by a factor 2 from that used in other situations, i.e. $T(\text{PWA}) = T/2$.) The functions P_L^0 are orthogonal when integrated over solid angle:

$$\int P_L^{0*}(\cos\theta) P_\lambda^0(\cos\theta) \, d\Omega = 4\pi\delta_{L\lambda}/(2L+1),$$

where $\delta_{L\lambda} = 1$ if $L = \lambda$ and 0 otherwise. The total elastic cross-section is

$$\sigma_{\text{EL}} = (4\pi/k^2) \sum_L (2L+1)|T_L|^2$$

$$= (\pi/k^2) \sum_L (2L+1)|1 - \eta_L e^{2i\delta_L}|^2.$$

The total inelastic cross-section is determined by the change in flux of outgoing beam particles when the scatterer is brought into place. This flux change for a solid angle $d\Omega$ is

$$(v/k^2)\left\{\left|\sum_L (2L+1)P_L^0(\cos\theta)\right|^2 - \left|\sum_L (2L+1)\eta_L e^{2i\delta_L} P_L^0(\cos\theta)\right|^2\right\} d\Omega.$$

Integration over all directions yields

$$(\pi v/k^2) \sum_L (2L+1)(1 - \eta_L^2).$$

The total inelastic cross-section (or reaction cross-section) is obtained by dividing the flux by the incident flux v, i.e.

$$\sigma_R = (\pi/k^2) \sum_L (2L+1)(1-\eta_L^2).$$

Finally, we add the elastic and inelastic cross-sections to give the total cross-section

$$\sigma_T = (2\pi/k^2)\Sigma(2L+1)(1-\eta_L \cos 2\delta_L).$$

A comparison between this expression and the expression for the elastic scattering amplitude shows that

$$\sigma_T = (4\pi/k)\mathrm{Im}\, f(0).$$

This is the Optical Theorem and is quite generally true. It is unusual in that it relates an intensity (σ_T) to an amplitude, in this case the imaginary part of the amplitude for elastic scattering in the forward direction.

The changes necessary for the case of meson-baryon scattering ($0^-\frac{1}{2}^+ \rightarrow 0^-\frac{1}{2}^+$) are small. For example:

$$\sigma_{EL} = (4\pi/k^2) \sum_{J,L} (J+\tfrac{1}{2})|T_{LJ}|^2,$$

where there are two partial waves for each given value of orbital angular momentum (L). These have angular momentum (J) equal to $L \pm \frac{1}{2}$. If we take the simplest case of the above formalism with only one partial wave active and this being elastic ($\eta_L = 1$), then

$$T = \sin \delta_L e^{i\delta_L} = 1/(\varepsilon - i),$$

where ε is cot δ_L. The end point of T drawn on an Argand diagram describes a path lying on a circle of radius $\frac{1}{2}$ centred at $i/2$. Figure 10.8(a) illustrates this feature. The circle is called the Unitarity circle. With $\eta_L < 1$ the end point would lie inside the circle and give a smaller elastic scattering cross-section. Any larger amplitude extending T outside the circle would give a scattering probability greater than unity, which is physically impossible.

When the phase shift δ_L goes through 90° the amplitude resonates. Then T and σ_{EL} go through a maximum. Near resonance we may use a Breit-Wigner form (Appendix B):

$$T = (\Gamma/2)/[(W_0 - W) - i\Gamma/2],$$

where W is the CM energy, W_0 the mass of the resonant state and Γ its width. The parameter ε is then given by

$$\varepsilon = 2(W_0 - W)/\Gamma.$$

When $W < W_0$, ε is positive and when $W > W_0$, ε is negative, so that the end point of T moves around the unitarity circle in an *anticlockwise* sense when going through a resonance.

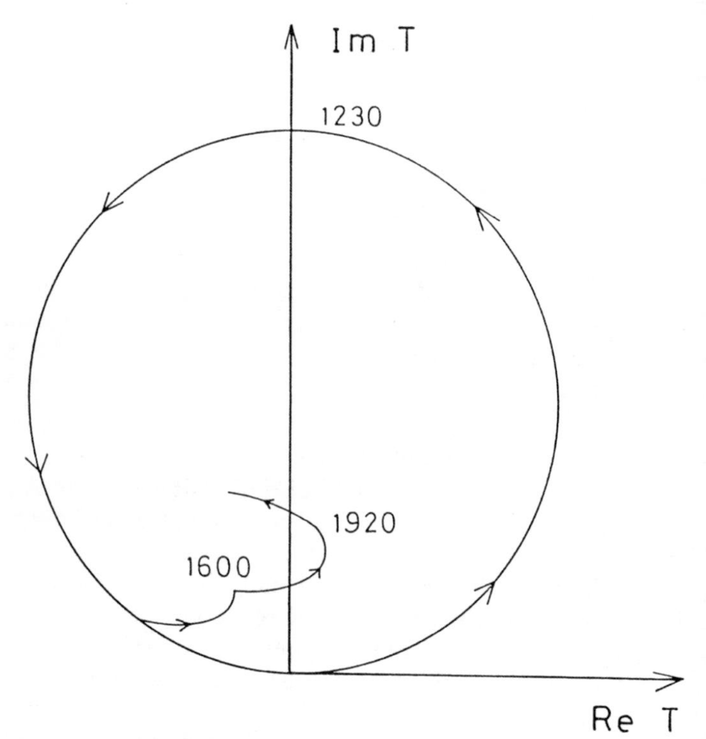

Fig. 10.8 (a) The Argand diagram for an elastic partial-wave $^{+}$p amplitude.

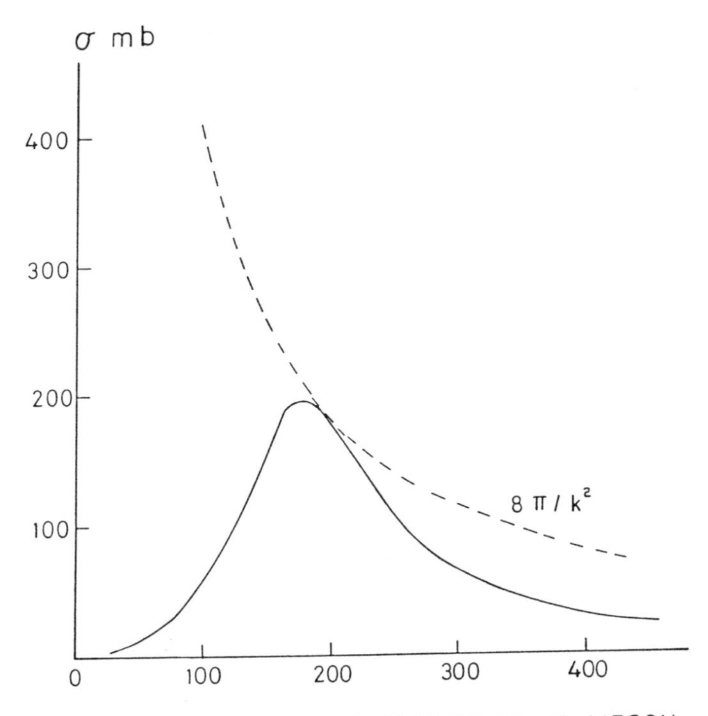

Fig. 10.8 (b) The energy dependence of the cross-section for π^+p elastic scattering near to the $\Delta^{++}(1232)$ resonance.

A rare example of a purely elastic resonance is provided by the $\Delta(1232)$ produced in π-nucleon scattering:

$$\pi^+ + p \rightarrow \Delta^{++}(1232) \rightarrow \pi^+ + p.$$

The resonance is elastic because the mass of 1232 MeV/c^2 is insufficient to allow a decay involving a second π-meson. Figure 10.8(b) illustrates the variation of cross-section with CM energy: the Breit-Wigner peak is centred at a CM energy of 1232 MeV. The π-nucleon cross-sections show a similar peak in all of the charge states Δ^{++}, Δ^+, Δ^0, Δ^-; from which it is concluded that the isospin of the $\Delta(1232)$ is $\frac{3}{2}$. The angular distribution of elastic scattering at resonance is also quite simple:

$$d\sigma/d\Omega \propto (1+3\cos^2\theta),$$

so that the scattering is purely in the *P*-wave. This limits the spin of the $\Delta(1232)$ to be either $\frac{1}{2}$ or $\frac{3}{2}$. However, the cross-section for a single-channel resonance is

$$\sigma_{EL} = 2\pi(2J+1)|T|^2/k^2.$$

At the resonance peak, T is i exactly, giving

$$\sigma_{EL} = 4\pi/k^2 \text{ for } J=\tfrac{1}{2}$$

and

$$\sigma_{EL} = 8\pi/k^2 \text{ for } J=\tfrac{3}{2}.$$

The measured cross-section is very close to $8\pi/k^2$ so that the spin parity of the $\Delta(1232)$ has to be $\frac{3}{2}^+$. On Fig. 10.8(a) the behaviour of the partial-wave amplitude for the ($I=\frac{3}{2}$, $J=\frac{3}{2}$) is followed through the $\Delta(1232)$ resonance and two higher mass and less elastic Δ resonances.

Analysis of baryon-formation experiments is normally more complicated than for the case of the $\Delta(1232)$. There can be several partial waves active, the resonances are usually only partly elastic and these resonances may lie on a background formed by the tails of other resonances. The technique has been refined so that now more than ten isospin $\frac{1}{2}$ (N) and a similar number of isospin $\frac{3}{2}$ (Δ) non-strange baryon resonances have been identified. Among the N's are examples with spin-parities $\frac{1}{2}^+$, $\frac{3}{2}^-$, $\frac{5}{2}^+$, $\frac{7}{2}^-$, $\frac{9}{2}^+$ and $\frac{11}{2}^-$ and masses 939, 1520, 1680, 2190, 2220 and 2600 MeV/c^2 respectively. Such states are obtained by increasing the relative orbital angular momentum of the quarks in integral steps. Partial-wave analysis has also been applied extensively to K-meson-nucleon elastic scattering angular distributions and to processes such as $\pi^+ p \rightarrow \pi^+ \Delta^+$.

Chapter 11
Hadron-hadron interactions

Hadron-hadron interactions can range in complexity from elastic scattering to processes, at the highest CM energies currently available, in which a hundred or more hadrons are emitted. The emerging hadrons have momenta transverse to the beam direction of typically 500 MeV/c or less. It follows that the QCD coupling strength at a typical vertex, α_s, is of order unity. As was explained in Chapter 6, a perturbative calculation of cross-section is not feasible in these circumstances.

A general qualitative account of hadron-hadron interactions is undertaken in section 11.1 employing geometric ideas. This approach makes use of the simple observation that in high-energy collisions the incident hadrons, as seen in the CM frame, are Lorentz contracted along the direction of motion into thin discs.

The only hadron-hadron cross-sections which can be reliably calculated using QCD are those for processes in which a parton from one hadron scatters off a parton from the other hadron with a large four-momentum transfer. In such cases the lowest order diagrams alone are important. After the 'hard' scattering the two partons each fragment into a 'jet' of collimated hadrons. These hard processes increase in importance with rising CM energy. They are discussed in section 11.2. At CM energies of a few GeV two-body processes of a different type are common:

$$\text{hadron} + \text{nucleon} \rightarrow \text{hadron} + \text{baryon}.$$

These are of theoretical interest because they require the exchange of some colour singlet, i.e. a virtual hadron. A hadron is a large fragile structure compared to a parton and so the likelihood of such processes falls off extremely rapidly with increasing four-momentum transfer or CM energy. They are discussed in section 11.3.

11.1 Geometric effects

At low CM energies the total cross-sections for hadrons on nucleons show rapid variations with energy due to the formation of resonant states like the $\Delta(1232)$ observed in π-nucleon scattering. This type of resonant structure dies away as the CM energy increases above a few GeV (Fig. 11.1).

Inferences about the relative magnitude of the total cross-sections can be made using the optical theorem and the quark model. The optical theorem (section 10.4) has

$$\sigma_T = 4\pi \mathrm{Im}\, f(0)/k,$$

where $f(0)$ is the forward elastic scattering amplitude. If the contributions to $f(0)$ of the various parton-parton scatterings add linearly, then

$$\sigma_T = (4\pi/k) \sum_{ij} \mathrm{Im}\, f_{ij}(0),$$

where $i(j)$ refers to a parton in the beam (target). In forward scattering the four-momentum transfer between the partons is small, so it is a fair approximation to limit consideration to valence-quark contributions. Then it follows that $\sigma_T(\pi p)$ should be about $(2/3)\sigma_T(pp)$: the experimental ratio is approaching this value as the CM energy increases. According to the same argument the ratio between particle and antiparticle cross-sections is expected to be unity. In fact there are big differences at low energies between $\sigma_T(\bar{p}p)$ and $\sigma_T(pp)$. The reason for this discrepancy is that the valence antiquarks in a beam antiproton can annihilate on a valence quark from a target proton, whereas this possibility is denied to valence quarks in a beam proton. However, as beam energies increase the q^2 involved in annihilations increases; the effect of increasing q^2 is that the importance of the valence quarks declines as they slowly submerge in the sea (e.g. see Fig. 9.4). Thus, it is to be expected

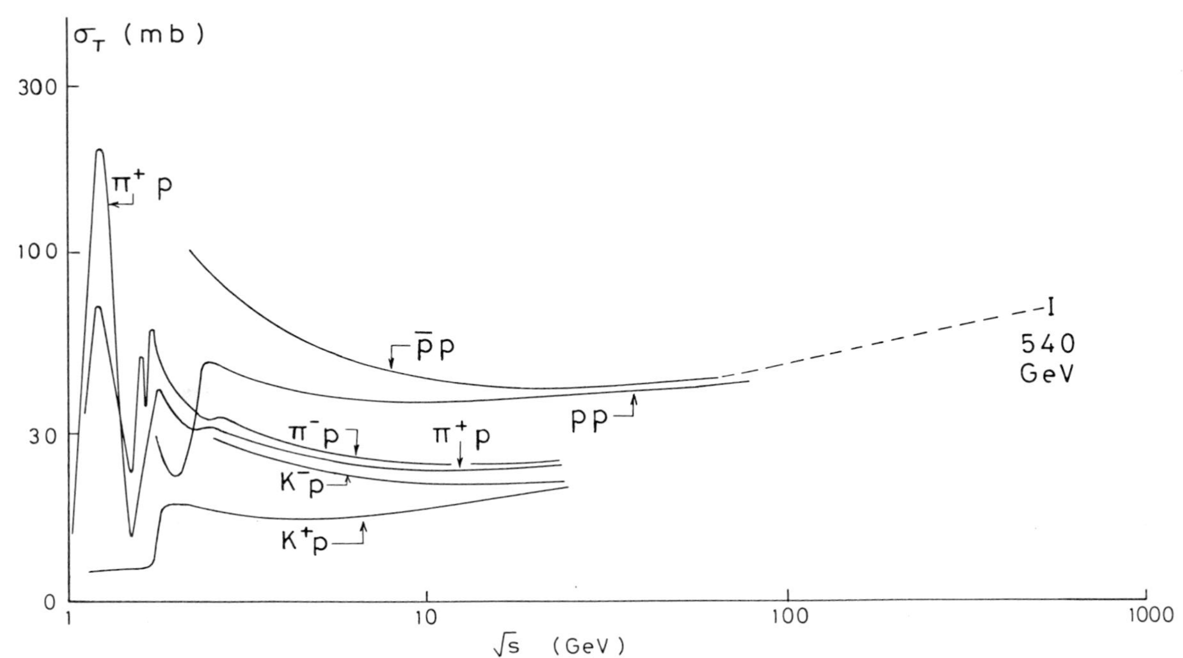

Fig. 11.1 Total cross-sections for hadron-hadron interactions as a function of total CM energy.

that particle and antiparticle total cross-sections converge with increasing CM energy. This expectation is borne out by the data shown in Fig. 11.1. A similar argument explains why the π-meson and K-meson cross-sections are also converging with increasing energy.

The elastic cross-section makes up about 15 per cent of the total cross-section and follows a similar energy variation. Elastic scattering shows a diffractive forward peak characteristic of scattering from an absorbing disc:

$$d\sigma/dq^2 = A\exp(bq^2),$$

where q^2 is the four-momentum transfer squared (recall q^2 is negative). We can relate the parameter b to the hadron size as follows. Suppose the hadron has a Gaussian shape:

$$U(r) \propto \beta e^{-r^2/R^2},$$

where β is an absorption coefficient. The amplitude for elastic scattering is given by

$$T = \int U(r)\exp[i(\mathbf{k}-\mathbf{k}')\cdot\mathbf{r}]\mathrm{d(volume)},$$

where $\mathbf{k}$ and $\mathbf{k}'$ are respectively the momenta of the incident and scattered beam particle. If the approximation is made that the target hadron is Lorentz contracted to a thin disc, this expression becomes

$$T = \int U(r)\exp(i\mathbf{Q}\cdot\mathbf{r})\,\mathrm{d(area)},$$

where $\mathbf{Q}$ is the transverse momentum of the scattered particle, and $Q^2 = -q^2$. Then

$$T = \beta\exp(-Q^2R^2/4)\int\exp[-(r/R - iQR/2)^2]\mathrm{d(area)}$$
$$\propto \exp(-Q^2R^2/4).$$

In turn, the differential cross-section is given by

$$d\sigma = |T|^2 2\pi Q\,dQ$$
$$\therefore d\sigma/dq^2 \propto \exp(q^2R^2/2).$$

We see that the slope exponent b is $R^2/2$. More detailed consideration shows that the elastic and total cross-sections are:

$$\sigma_{EL} = \pi\beta^2 R^2/2 \quad \text{and} \quad \sigma_T = 2\pi\beta R^2.$$

For proton-proton scattering at 52·8 GeV CM energy σ_T is measured to be 42·5 mb, σ_{EL}/σ_T is 0·18, and the slope $R^2/2$ is around 11 GeV^{-2}. These values yield $R \sim 1{\cdot}0$ fermi and $\beta \sim 0{\cdot}7$, so the proton is behaving as a very absorbing disc which is of similar radius to that measured by *elastically* scattering electrons from protons.

If the cross-section for single-particle production is measured at a given angle and momentum ignoring the presence (or absence) of other particles the result is called an *inclusive* cross-section. The covariant form for this cross-section is

$$E\frac{d^3\sigma}{dp^3} = E\frac{d^2\sigma}{dp_T^2\, dp_L},$$

where p_L and p_T are the longitudinal and transverse momentum components and E is the energy (here dp_T^2 means $2\pi p_T dp_T$). Measurements made for all outgoing hadrons (which are predominantly mesons) fit the form

$$E d\sigma/dp_T^2\, dp_L \propto \exp(-Ap_T)$$

and, for values of p_T below 1 GeV/c, the value of A is typically around 6. This implies that the transverse dimension of the hadron interaction region, which is of order $1/\langle p_T \rangle$, is $6(\mathrm{GeV}/c)^{-1}$ or 1·2 fermi, in agreement with the previous result.

Identical bosons emitted from a hadron-hadron interaction exhibit Bose-Einstein correlations: the effect is an enhanced probability for finding them with small relative momenta. Hanbury-Brown and Twiss (1954, 1956) exploited the analogous effect for photons when making measurements of stellar sizes, and Cocconi (1974) amongst others proposed an equivalent measurement of the size of hadronic interactions.

Suppose that identical bosons are emitted from sources S_1 and S_2 in a hadronic interaction and that the coincidence is recorded with detectors D_1 and D_2 (Fig. 11.2(a)). Now the case

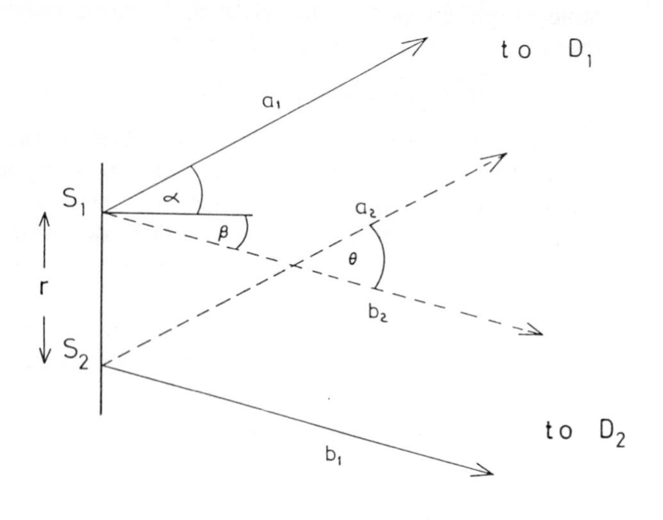

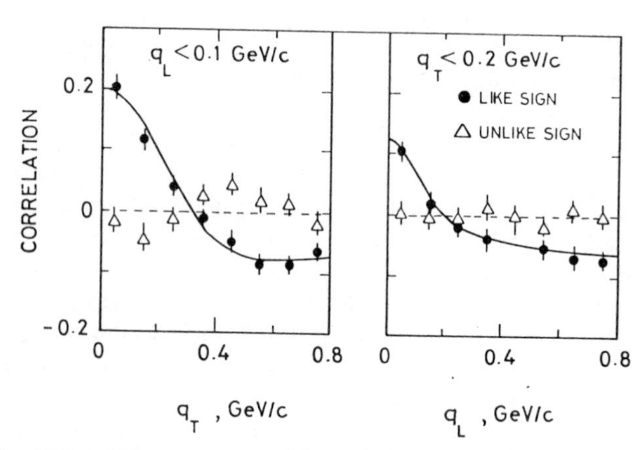

Fig. 11.2 (a) The geometry of the emission of identical bosons at S_1 and S_2 and their detection at D_1 and D_2. (b) Pair correlations for like-sign and unlike-sign mesons emitted from pp interactions at 50 GeV CM energy (Drijard *et al.*, *Nucl. Phys.* **B155**, 269 (1979)).

in which S_1 emits boson 1 and S_2 emits boson 2 is indistinguishable from the case in which S_1 emits boson 2 and S_2 emits boson 1. The amplitudes therefore add coherently, giving

$$A=\exp(-[q_1a_1+q_2b_1])+\exp(-[q_1a_2+q_2b_2]),$$

where q_1 and q_2 are the boson three-momenta and a_1, a_2, b_1 and b_2 are the distances between source and detector indicated in Fig. 11.2(a). Thus the probability is

$$I\propto|A|^2=|1+\exp(i[q_1\Delta a+q_2\Delta b])|^2,$$

where $\Delta a=a_1-a_2$, $\Delta b=b_1-b_2$. If the boson momenta q_1 and q_2 are nearly equal,

$$q_1\Delta a+q_2\Delta b=qr(\alpha+\beta)=qr\theta,$$

where r is the separation between S_1 and S_2 while θ is the angle subtended at either S_1 or S_2 by D_1D_2.

$$\therefore\ I(\theta)\propto 1+\cos(qr\theta).$$

Measurements of the coincidence rate as a function of the angle θ between the detectors would show maxima when $qr\theta$ is $2\pi, 4\pi$, etc. Hence r could be determined. Integration over the whole interaction region shows that the coincidence rate is enhanced by a factor:

$$\rho_{12}=1+\lambda'[2J_1(q_TR)/q_TR]^2,$$

where J_1 is the first-order Bessel function; q_T and q_L are the components of $(\mathbf{q}_1-\mathbf{q}_2)$ taken respectively perpendicular and parallel to the symmetry axis $(\mathbf{q}_1+\mathbf{q}_2)$. R is the radius of the region over which the radiators are uniformly distributed. λ' contains a damping term dependent on the lifetime τ of the sources

$$\lambda'=\lambda/[1+(q_L\tau)^2]$$

and λ takes account of incomplete interference $(0\leqslant\lambda\leqslant 1)$. Figure 11.2(b) shows that correlations are present for like-sign and absent for unlike-sign π-mesons emitted from the collisions of protons at 50 GeV CM energy (Drijard *et al.* (1979)). Best fits to this data yield a radius of $1{\cdot}34 \pm 0{\cdot}31$ fermis for the interaction region.

Feynman (1969) noted that in the longitudinal dimension (along beam axis $0z$) colliding hadrons appear Lorentz contracted to delta functions $\delta(z)$ in the high-energy limit. The delta function $\delta(z)$ is a distribution which has a spike at $z=0$ and unit area. For other values of z it vanishes. One property of the delta function is that

$$\int f(z)\delta(z)\,\mathrm{d}z = f(0)$$

provided the range of integration contains $z=0$. In order to obtain the longitudinal momentum distribution of the partons emitted we Fourier transform the shape of the active region:

$$\begin{aligned} f(p_{\mathrm{L}}) &= \int_{-\infty}^{+\infty} \delta(z)\exp(ip_{\mathrm{L}}z)\,\mathrm{d}z \\ &= 1, \end{aligned}$$

giving a flat distribution in p_{L}. These partons reorganize themselves into hadrons which share out the flat longitudinal momentum distribution. Therefore the observed *hadron* longitudinal momentum distribution is

$$N(p_{\mathrm{L}})\,\mathrm{d}p_{\mathrm{L}} \propto \mathrm{d}p_{\mathrm{L}}/E,$$

where E is the hadron energy. At sufficiently large CM energy

$$E = (m^2 + p_{\mathrm{T}}^2 + p_{\mathrm{L}}^2)^{\frac{1}{2}} \simeq p_{\mathrm{L}},$$

so that in terms of the Feynman x-variable ($x = p_{\mathrm{L}}/p_{\mathrm{L}}(\max)$)

$$N(x)\,\mathrm{d}x \propto \mathrm{d}x/x.$$

Another variable used to describe longitudinal momenta is the rapidity

$$y = \tfrac{1}{2}\ln\left[\frac{E+p_{\mathrm{L}}}{E-p_{\mathrm{L}}}\right] = \tanh^{-1}(p_{\mathrm{L}}/E),$$

where β is the particle velocity. The rapidity has the property that under a boost $\beta'(=\tanh\omega)$ along the z-axis

$$y \rightarrow y' = y + \omega.$$

Rapidity distributions therefore look the same shape whether viewed in the CM, beam or target rest frames. The maximum value of rapidity is achieved when the final state contains just two particles; one travelling along the beam direction in the CM frame, the other travelling in the opposite direction. Then at high enough energies,

$$E(\max) \simeq \sqrt{s}/2,$$
$$E(\max) + p_{\mathrm{L}}(\max) \simeq \sqrt{s},$$
$$E^2(\max) - p_{\mathrm{L}}^2(\max) = \mathrm{m}^2.$$

Combining these results the maximum rapidity is:

$$y^*(\max) = \ln(\sqrt{s}/m).$$

In terms of y the single-particle distribution deduced above becomes

$$N(y)\,\mathrm{d}y = \text{constant} \times \mathrm{d}y.$$

Feynman combined these ideas with the idea that the transverse momenta of the hadrons emitted from an interaction are determined by the incident particle's size and opacity. Feynman's scaling hypothesis can now be stated for the spectra of particles emitted from hadron-hadron collisions: single-particle spectra depend at asymptotically high energies only on p_{T} and x (or y), i.e.

$$\underset{s \rightarrow \infty}{\mathrm{Lt}}\; E\left(\frac{\mathrm{d}^3\sigma}{\mathrm{d}p^3}\right) = f(x, p_{\mathrm{T}}).$$

Figure 11.3 shows the experimentally observed rapidity distributions of hadrons from p-p and $\bar{\mathrm{p}}$-p interactions. Evidently a plateau in y does develop in the central region but its height increases logarithmically with s. This increase coupled with the increase in the rapidity range ($y^*(\max) \sim \ln s$) means that the mean multiplicity $\langle n \rangle$ shows a component which increases like

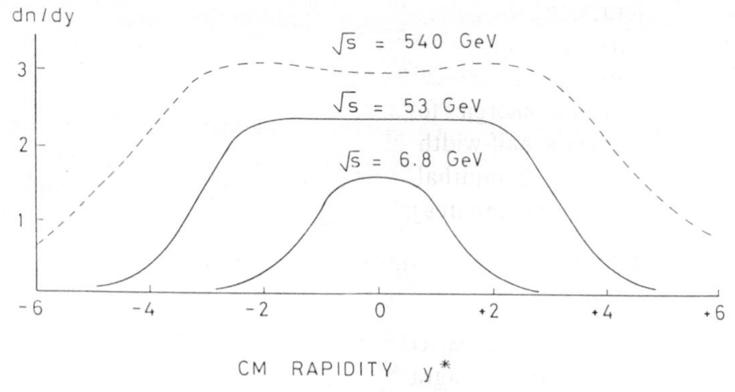

Fig. 11.3 The rapidity (y) distributions for particles emitted in p-p and p̄-p interactions (excluding diffractive processes).

$(\ln s)^2$. For charged particles the measured mean multiplicity varies, for s in MeV^2, like:

$$\langle n \rangle = 0{\cdot}57 + 0{\cdot}584 \ln s + 0{\cdot}127 (\ln s)^2$$

reaching 29 at 540 GeV CM energy. Weak violations of Feynman scaling are also observed in the rise in mean transverse momentum of secondary hadrons; from 0·35 to 0·42 GeV/c as the CM energy climbs from 50 to 540 GeV. Over the same range the slope parameter b of the forward elastic peak changes from 13 to 18 $(\mathrm{GeV}/c^2)^2$ as the peak sharpens. Feynman's intuitive insight is broadly correct with only logarithmic violations seen to scaling. This is very valuable because QCD calculations of gross features of hadron production are quite opaque. This criticism does not, however, apply to the calculation of hard parton-parton scattering leading to jets of hadrons. Jet production therefore forms the subject of the next section.

11.2 Jet production

The cross-section for recognizable jets is of order 1 μb in the 540 GeV p̄-p collisions observed at the CERN SPS p̄p collider. The energies deposited in cells of the calorimeters around the interaction region (e.g. see Fig. 8.5) constitute the raw measurements. Energy depositions associated into a jet are treated like momenta and added vectorially to give 'longitudinal' and 'transverse' energies for a jet (E_L, E_T). The energy deposition in a jet has a half-width of about one unit of rapidity and one radian in azimuthal angle (ϕ) around the beam axis. Figure 11.4 compares jet cross-sections at 540 and 60 GeV CM energy.

The data on two jets taken at 540 GeV has already been presented in Fig. 6.5 and shows the angular dependence $(1-\cos\theta^*)^{-2}$ expected if single-gluon exchange is dominant. Some Feynman diagrams for the lowest order QCD contributions are shown as insets in Fig. 11.4: of these 11.4(a) to 11.4(c) are the single-gluon exchange diagrams. The total process

$$\text{hadron} + \text{hadron} \rightarrow \text{jet} + \text{jet}$$

requires the more complicated diagram shown in Fig. 11.5. In this diagram the process is factorized into distinct steps: partons emerge from the parent hadrons then follows a hard parton-parton scattering and this is followed by independent fragmentation of these two partons. Factorization makes it straightforward to write the differential jet cross-section

$$\mathrm{d}^2\sigma/\mathrm{d}x_1\mathrm{d}x_2\,\mathrm{d}(\cos\theta^*)$$
$$=\sum_{ij}\frac{F_i(x_1)}{x_1}\frac{F_j(x_2)}{x_2}K\frac{\mathrm{d}\sigma_{ij}}{\mathrm{d}(\cos\theta^*)},$$

where $\mathrm{d}\sigma_{ij}/\mathrm{d}(\cos\theta^*)$ is the cross-section for a parton of species i (quark, gluon, antiquark) to scatter from a parton of species j. θ^* is the scattering angle in the parton-parton CM frame. $F_i(x_1)\,\mathrm{d}x_1/x_1$ is the number of quarks of species i carrying a longitudinal momentum fraction between x_1 and $x_1+\mathrm{d}x_1$ in the first hadron. $F_j(x_2)\,\mathrm{d}x_2/x_2$ has a corresponding significance for quarks of species j in the second hadron. There is a factor unity for the parton fragmentation into isolated jets. A major assumption in factorization is that the structure functions $F_i(x_1)$ and $F_j(x_2)$ are precisely those measured in deep inelastic lepton scattering. A test of this facet of factorization has already been

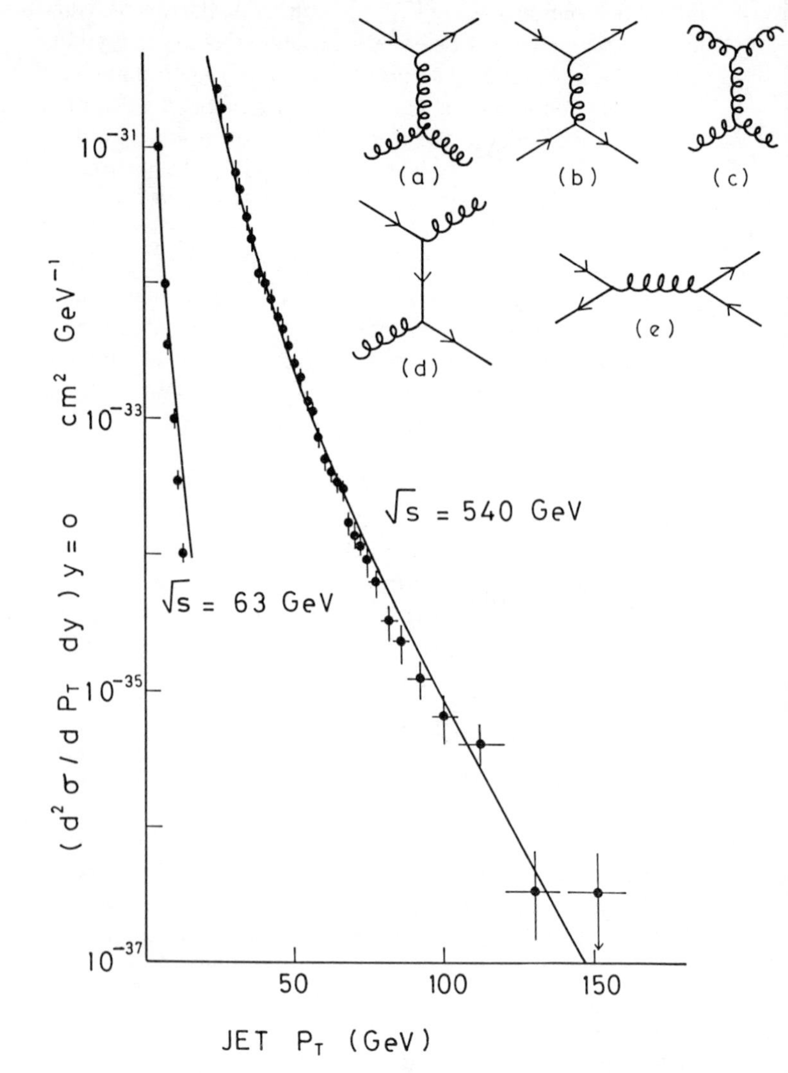

Fig. 11.4 The measured jet differential cross-sections as a function of jet transverse energy at 63 and 540 GeV CM energy (after Jacob, 1983). Insets show examples of the contributing first-order QCD diagrams.

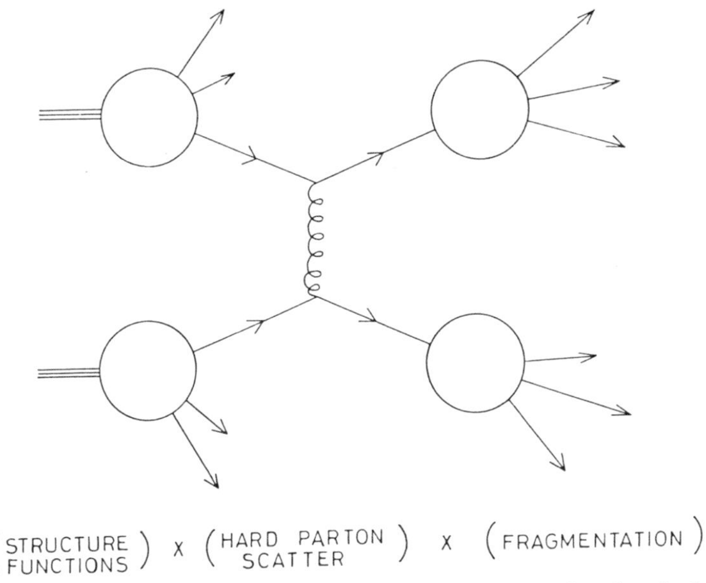

Fig. 11.5 The factorized process for hadron + hadron→jet + jet + hadrons.

provided by studies of the Drell-Yan process (section 9.4). The K factor used there must also be incorporated here in the jet cross-section so as to correct for soft gluon emission. This analysis has been applied to two-jet production in $\bar{p}p$ collision at 540 GeV CM energy by Arnison *et al.* (1984). The authors extract a structure function which is found to agree well with predictions based on structure functions measured in ν-scattering experiments. Evidently the ideas on factorization which underlie the calculations made here are working well. The QCD-based predictions for jet differential cross-sections are superposed on Fig. 11.4.

The way that partons fragment into hadrons has been studied in e^+e^- collisions and at the $\bar{p}p$ collider. In the case of e^+e^- collisions measurements are simpler to interpret because when there are two jets they are both quark jets and there are no other partons. Figure 11.6 shows the distribution in z, the component

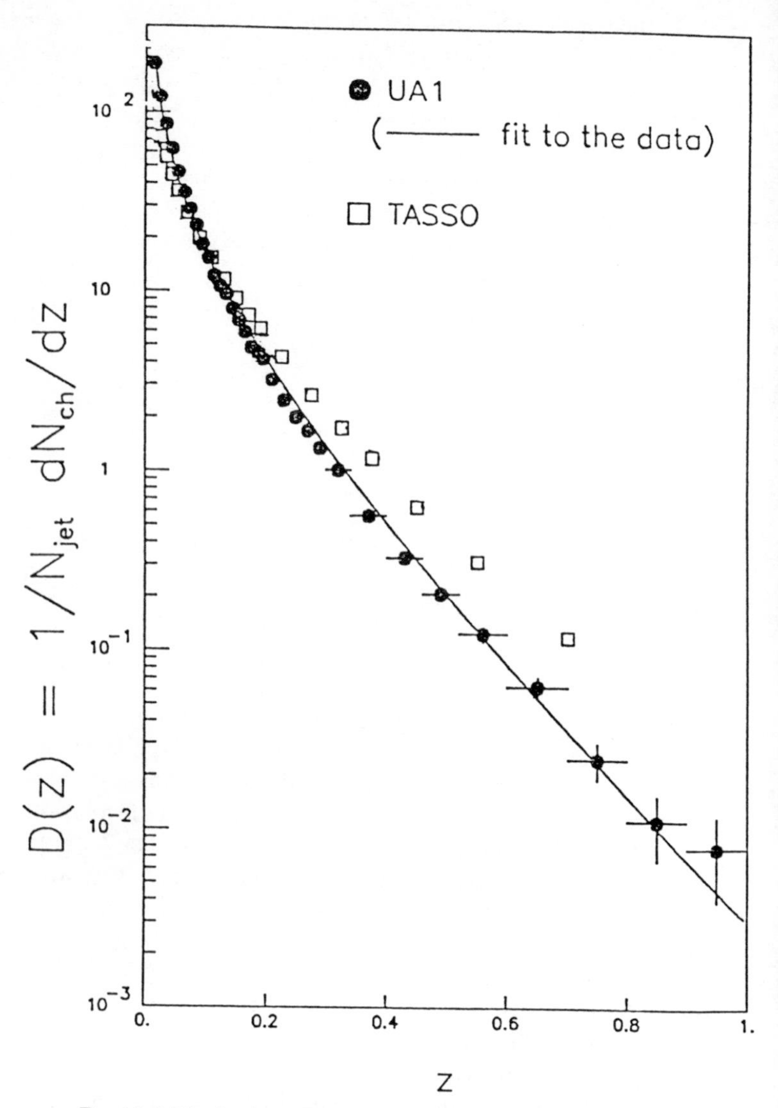

Fig. 11.6 The jet fragmentation functions measured for jets produced in e^+e^- (open squares) and $\bar{p}p$ (solid circles) collisions (after Arnison *et al.*, CERN-EP/86-55 (May 1986)).

of the parton longitudinal momentum carried by an individual hadron. The comparison is made between jets produced in e^+e^- and $\bar{p}p$ collisions. The z-distribution for the $\bar{p}p$ jets is peaked at lower z-values for two reasons. First, the parton-parton CM energy is larger (95 GeV against 34 GeV) so q^2 values are larger. Second, the $\bar{p}p$ jets are dominantly gluon jets while those from e^+e^- collisions have quark parents. The transverse momentum (p_T) spectrum of the hadrons with respect to the jet axis falls exponentially with increasing p_T: the mean value is around 0·5 GeV/c. QCD-based predictions fit these observed longitudinal and transverse momentum distributions of single particles in jets. Three-jet production has been discussed earlier in Chapter 6 in connection with a measurement of α_s.

11.3 Peripheralism and Reggeized exchange

The differential cross-sections for the two hadron to two hadron reactions fall off very rapidly away from the forward direction, provided the measurement is made well above the threshold energy. Figure 11.7 shows the cross-section as a function of $t(=q^2$ the four-momentum transfer squared) for

$$K^+ + p \rightarrow K^{*0} + \Delta^{++}$$

at several beam momenta from a compilation by Ciapetti (1973). Small numerical values of t imply large separations hence such processes are called peripheral. Another characteristic is that the total cross-section for producing a two-body final state falls off with energy like

$$\sigma \sim s^{-n},$$

where n lies in the range from 1 to 4. At low energies two-body peripheral processes have cross-sections of order 10 μb but they have fallen off to insignificant levels at collider energies (540 GeV). These peripheral processes evidently involve more than single-parton exchange; in fact exchange of a colour-singlet cluster of partons is required. In the example given above

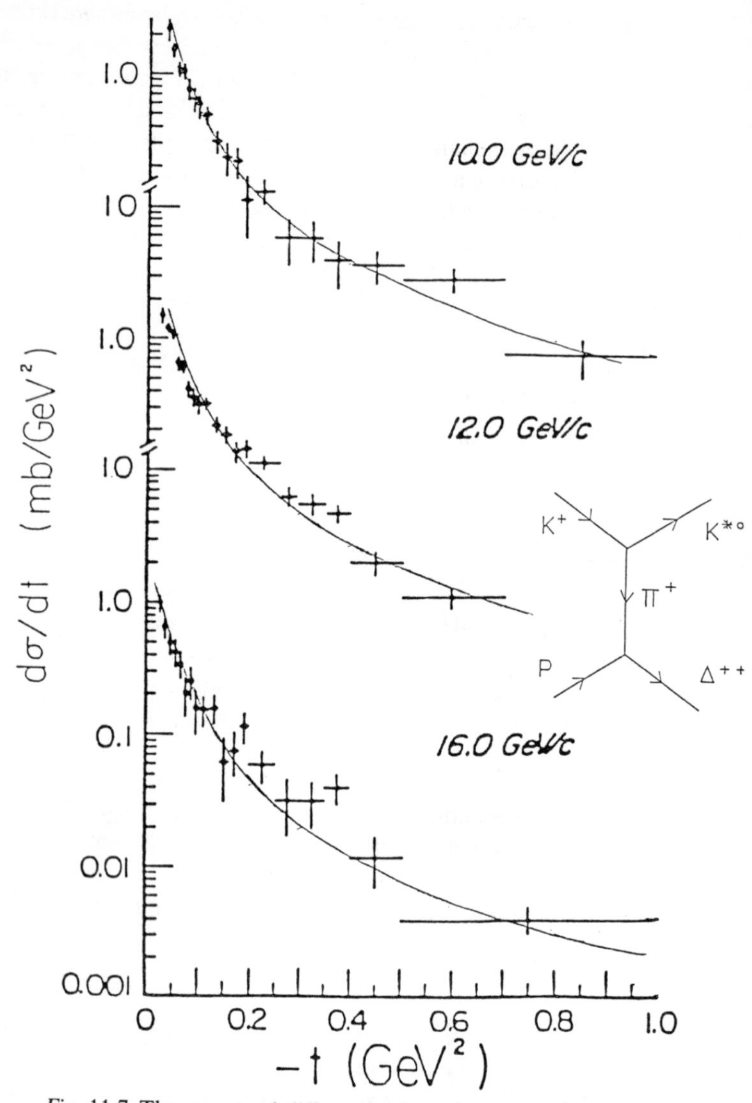

Fig. 11.7 The measured differential cross-section for $K^+ + p \rightarrow K^{*0} + \Delta^{++}$ as a function of $-t$ $(-q^2)$ (after Ciapetta *et al.* (1973)). The curves indicate the variation expected for the contribution from the π-meson exchange diagram shown in the inset.

the emerging hadrons couple naturally to π-mesons; their decays are

$$K^{*0} \rightarrow K^{+} + \pi^{-} \quad \text{and} \quad \Delta^{++} \rightarrow p + \pi^{+}.$$

The most likely exchange, therefore, is that drawn as an inset in Fig. 11.7, involving a virtual π-meson. If we follow uncritically the techniques of field theory to write the amplitude corresponding to this diagram we obtain

$$T = \text{vertex factors}/\pi\text{-meson propagator}$$

so that

$$d\sigma/dt \propto (m_{\pi}^{2} - t)^{-2}.$$

Superposing this shape on Fig. 11.7 is seen to provide an excellent fit to the data. This and other successes make it at least plausible that virtual hadrons are exchanged in peripheral processes. In the reaction $1 + 2 \rightarrow 3 + 4$:

$$\begin{aligned} t = q^{2} &= (E_{1} - E_{3})^{2} - (\mathbf{p}_{1} - \mathbf{p}_{3})^{2} \\ &= m_{1}^{2} + m_{3}^{2} - 2(E_{1}E_{3} - \mathbf{p}_{1} \cdot \mathbf{p}_{3}). \end{aligned}$$

This has an extreme value in the forward direction:

$$t_{\min} = m_{1}^{2} + m_{3}^{2} - 2(E_{1}E_{3} - p_{1}p_{3}).$$

When the scattering is elastic $t_{\min}$ is zero but generally it is negative (see Appendix A). Thus $(t - m^{2})^{-2}$ is more sharply peaked in the forward direction if m, the mass of the exchanged hadron, is small. This agrees with the uncertainty principle: the lighter the exchanged mass of the particle, the longer the range of interaction and hence the more peripheral the process. The one-particle exchange model provides qualitative guidance but we now show that refinement is necessary if it is to be self-consistent. The reactions (s) $\pi^{+} + p \rightarrow \pi^{+} + p$, (t) $\bar{p} + p \rightarrow \pi^{-} + \pi^{+}$ and (u) $\pi^{-} + p \rightarrow p + \pi^{-}$ are related by 'crossing' particles from incoming to outgoing. Figure 11.8 shows this with an indication of how the Δ^{++} can contribute: as a resonance in (s) and as an exchange in (t) and (u). We now define the Lorentz-invariant quantities for reaction (s):

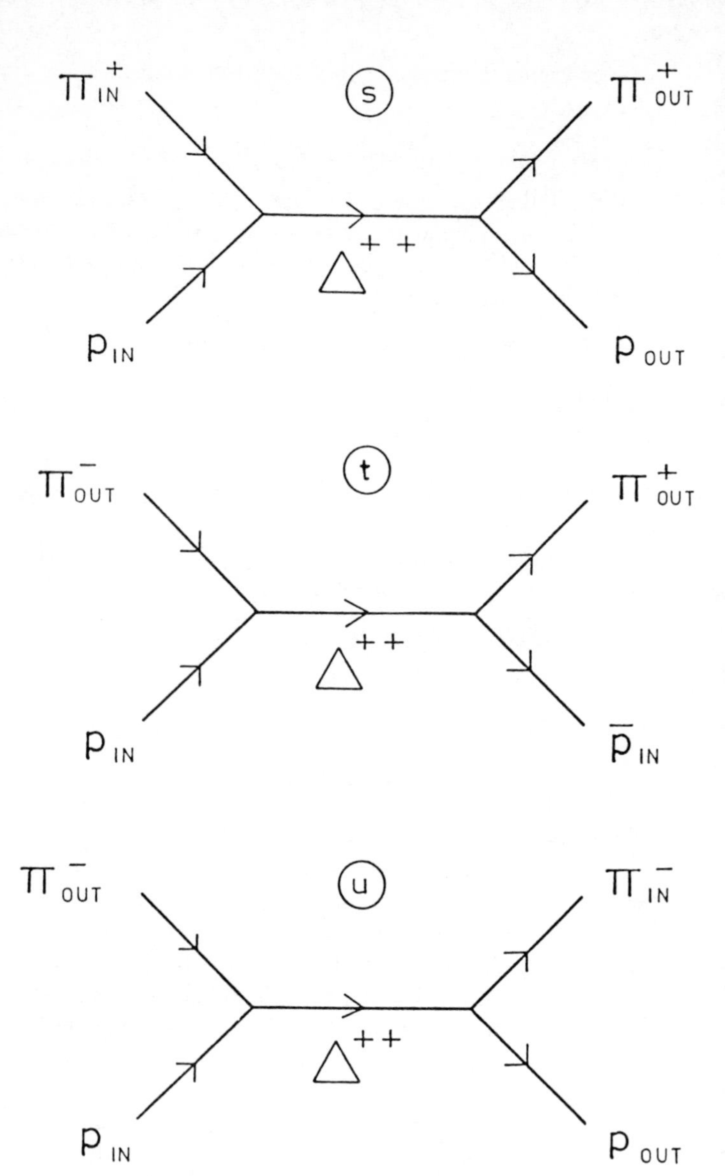

Fig. 11.8 Three reaction channels related by crossing: $\pi^+ + p \rightarrow \pi^+ + p$, $\bar{p} + p \rightarrow \pi^- + \pi^+$ and $\pi^- + p \rightarrow p + \pi^-$.

$$s(\text{CM energy squared}) = [p(\pi^+_{\text{in}}) + p(\text{p}_{\text{in}})]^2$$

$$t(\text{four-momentum transfer squared}) = [p(\text{p}_{\text{in}}) - p(\text{p}_{\text{out}})]^2$$

where $p(\text{p}_{\text{in}})$, etc., are all four-momenta. Correspondingly in reaction (t):

$$s_t = [p(\text{p}_{\text{in}}) + p(\bar{\text{p}}'_{\text{in}})]^2 = [p(\text{p}_{\text{in}}) - \text{p}(\text{p}'_{\text{out}})]^2 = t,$$

and

$$t_t = [p(\text{p}_{\text{in}}) - p(\pi^-_{\text{out}})]^2 = [p(\text{p}_{\text{in}}) + p(\pi^+_{\text{in}})]^2 = s,$$

where we have used the fact that an outgoing (ingoing) antiparticle of energy E and momentum $\mathbf{p}$ is equivalent to an ingoing (outgoing) particle of energy $-E$ and momentum $-\mathbf{p}$. The experimentally accessible region of the s-channel has $s > 0$, $t < 0$ and the experimentally accessible region for the t-channel has $s < 0$ and $t > 0$; they are completely disconnected kinematic regions. The u-channel is also disconnected from both the other channels.

Now we can return to consider the refinement of the one-particle exchange model. The incident and outgoing particles involved will be assumed spinless and the intermediate particle of spin l will be assumed to be a t-channel resonance of mass m. The four-momentum transfer in the t-channel is given at large CM energies by

$$t_t = -2p_t^2(1 - \cos\theta_t),$$

where p_t is the three-momentum of either incident hadron, and θ_t is the scattering angle. Then $s_t = 4p_t^2$ and

$$\begin{aligned}\cos\theta_t &= 1 + 2t_t/s_t \\ &= 1 + 2s/t.\end{aligned}$$

Thus, the amplitude for the propagation and decay of the resonance is

$$\begin{aligned}T &= g^2 P_l(\cos\theta_t)/(t - m^2) \\ &= g^2 P_l(1 + 2s/t)/(t - m^2).\end{aligned}$$

We now make use of the powerful idea that although the s- and t-channels are physically distinct processes the same amplitude applies to both because their diagrams are related by crossing. This would certainly be the case in QED. The amplitude must therefore be continued analytically outside the range $-1 \leqslant \cos\theta_t \leqslant +1$ defining the t-channel and into the s-channel. The result at large CM energy ($s \to \infty$) is that

$$T \approx P_l(2s/t+1) \to s^l$$

giving a divergent cross-section. This divergence has arisen through assigning a fixed angular momentum to the exchange. Why should a virtual particle with negative mass squared possess the same angular momentum as the real particle with positive mass squared? A general answer to the difficulty was provided by Regge (1959). He proposed that the angular momentum of an exchange has a linear dependence on t:

$$\alpha(t) = \alpha_0 + \alpha' t.$$

This relation defines what is called the particle trajectory as shown in Fig. 11.9. Then real bosons will occur whenever $\alpha(t)$ is exactly a positive integer (half-integer for fermions). If m^2 is the value of t for which the angular momentum reaches the integer l then

$$l = \alpha_0 + \alpha' m^2.$$

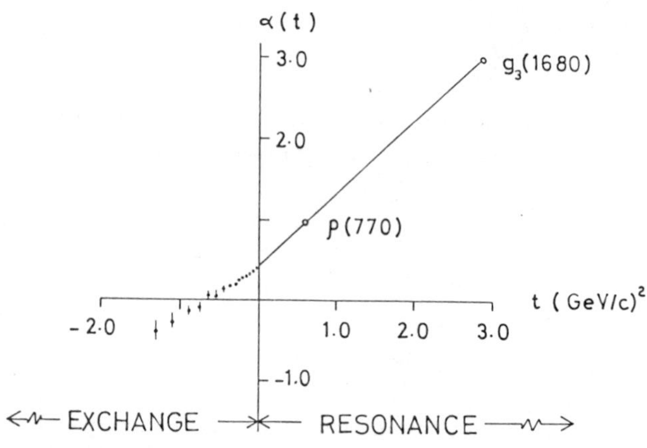

Fig. 11.9 The Chew-Frautschi plot for the Regge trajectory of the ρ-meson (after Barnes *et al.*, *Phys. Rev. Lett.* **37**, 76 (1976)).

Subtracting this from the general expression above gives

$$(m^2 - t) = (l - \alpha(t))/\alpha'.$$

The amplitude can now be written as a partial-wave expansion

$$T = \sum_l \beta(t) P_l(\cos\theta_t)/(l - \alpha(t))$$

where $\beta(t)$ specifies the strength of the coupling of the exchange at the vertices with an incoming and outgoing hadron. At large energies in the s-channel $P_l(\cos\theta_t)$ tends to $(\cos\theta_t)^l$ so that

$$T \rightarrow \beta(t)(\cos\theta_t)^{\alpha(t)} \propto \beta(t) s^{\alpha(t)}$$

at a fixed value of t. Finally the differential cross-section for the two-body final state at large s and fixed t is

$$\mathrm{d}\sigma/\mathrm{d}t \propto |T|^2/s^2 \propto F(t) s^{2\alpha(t)-2},$$

where $F(t)$ is a function of t alone. If several exchanges are possible then the exchange whose trajectory $\alpha(t)$ lies highest will dominate at large enough s. This behaviour has been verified in a large number of reactions (Irving and Worden (1977)). Here we consider the reaction

$$\pi^- + \mathrm{p} \rightarrow \pi^0 + \mathrm{n}$$

which is dominated by exchange of the ρ-meson trajectory. Data at beam momenta from 20 to 200 GeV/c is available and Barnes *et al.* (1976) have extracted the values of $\alpha(t)$ at several values of t using the equation

$$(\mathrm{d}\sigma/\mathrm{d}t)_{s'}/(\mathrm{d}\sigma/\mathrm{d}t)_s = (s'/s)^{2\alpha(t)-2}.$$

Figure 11.9 displays the extracted values of $\alpha(t)$ on what is called a Chew-Frautschi plot of $\alpha(t)$ against t. It is clear that the trajectory determined in the t-channel region lies along the continuation of a trajectory through the $\rho(1^-)$ and $g_3(3^-)$ resonances in the s-channel. Regge's analysis offers a consistent interpretation of the data though not at the level of a full field-theoretic interpretation. The g-meson is an angular momentum excitation of the ρ-meson: it is termed a recurrence of the ρ-meson. All the measured Regge trajectories are nearly parallel, with slopes of around 0·9 on the Chew-Frautschi plot. This common slope appears to be determined by the energy density per unit length (λ) in the gluon flux tube between the quarks.

We picture a meson as a pair of massless quarks connected by a string of length $2R$ which is rotating about the mid-point. Then the relativistic energy in the string is

$$M = 2\int_0^R \lambda\gamma \, \mathrm{d}r,$$

where $\gamma = (1-\beta^2)^{-\frac{1}{2}} = (1-r^2/R^2)^{-\frac{1}{2}}$, if the ends of the tube move with the velocity of light ($\beta = 1$ when $r = R$). Integrating this expression gives

$$M = \lambda R \pi.$$

Similarly, the orbital angular momentum:

$$L = 2\int_0^R \lambda\beta\gamma r \, \mathrm{d}r = \lambda R^2\pi/2.$$

This gives a slope $(2\lambda\pi)^{-1}$ for trajectories. The observed value of the slope is 0·9 $(\mathrm{GeV}/c)^{-2}$, and λ is thus 0·16. It is interesting that this same value for λ also emerges when potential model fits are made to quarkonia energy levels (see sections 6.4 and 10.1).

At high energy the contribution of a Regge exchange to the cross-section has an energy dependence $\sigma \sim s^{2\bar{\alpha}-2}$, where $\bar{\alpha}$ is the mean value of $\alpha(t)$. Angular distributions are peaked at $t=0$, so that $\bar{\alpha} \simeq \alpha(0)$. Now the intercepts $\alpha(0)$ of the Regge trajectories for mesons and baryons are all less than unity; thus two-body hadronic cross-sections must fall off with increasing CM energy like s^{-n}. Elastic scattering is a two-body to two-body process but its cross-section is anomolous in changing little with increasing CM energy above about 5 GeV. If a Regge interpretation of this feature is attempted the equivalent exchange has a trajectory with $\alpha(0)$ equal to unity. The exchange at work in elastic scattering is often called the Pomeron. None of the known mesons lies on this trajectory, and an appropriate view is that the exchange contributing is that of two or more gluons.

Chapter 12
Ways forward

The preceding chapters have been used to develop the theory of elementary particles known as the standard model. This model is built on the requirement that physics is invariant under local gauge transformations of the $SU(3)_c \otimes SU(2)_L \otimes U(1)$ symmetry groups. It is a highly successful model: the features of the electroweak interactions are calculable using perturbation theory; QCD can accommodate simultaneously the asymptotic freedom of quarks at large q_2 and their confinement at low q^2 into hadrons. The means for making realistic calculations for soft (low q^2) processes using QCD are being developed: one approach uses what is called lattice gauge theory (see Callaway (1985)). However, the standard model leaves many questions unanswered. A partial list is the following. Why are the charges of the electron and proton equal in magnitude? Why is nature left-handed? Why are there several generations of leptons and quarks (u, d, e^-, ν_e), (c, s, μ^-, ν_μ), (t, b, τ^-, ν_τ)? There is a surprising number of fundamental 'constants' whose origin is obscure: the masses of leptons and quarks, the elements of the Kobayashi-Maskawa mixing matrix, the weak, electromagnetic and strong coupling constants, the Weinberg angle and the mass (and even existence) of the Higgs boson. What, then, are the possibilities for improving our understanding? Some popular ideas are considered below, but the reader must remain aware that a complete theory has yet to emerge.

New accelerators now under construction or being designed should offer new experimental possibilities. A list of these is given in Table 12.1. There is a need for radical innovation in

Table 12.1

Accelerator	*Beam type and energy (GeV)*	*Date of commissioning*	*Location*
SPS collider	$\bar{p}(100\rightarrow450) + p(100\rightarrow450)$	1981	CERN (Geneva)
Fermilab collider	$\bar{p}(1000)+p(1000)$	1986	Fermilab (Illinois)
SLC	$e^-(50)+e^+(50)$	1987	SLAC (California)
LEP	$e^-(50)+e^+(50)$	1988	CERN (Geneva)
HERA	$e^-(30)+p(800)$	1988	DESY (Hamburg)
LEP II	$e^-(100)+e^+(100)$	1992?	CERN (Geneva)
SSC	p(20,000)+p(20,000)	1996	USA
LHC?	$\bar{p}(8000)+p(8000)$	?	CERN (Geneva)

accelerator design because the construction costs of synchrotrons and their energy demands during operation would make yet higher energies prohibitively expensive.

With high-energy e^+e^+ colliders the experimentalists will be able, by scanning the beam energy, to search for threshold effects associated with new flavours (a change in R, narrow resonances, three-jet decays of quarkonia). This could provide some evidence for the existence of the top quark. With the machine CM energy set to the Z^0 mass ($\times c^2$) annihilations will produce Z^0s at rates of 10^4/day. There will be the chance to examine the decay products for any signs of the Higgs meson or of heavier flavours with quark masses below ~ 46 GeV/c^2. There will also be the opportunity to scan the beam energy across the Z^0 and so measure its width directly. As explained in section 8.4 the number of light (including massless) neutrino species is easy to extract from the measured value of the Z^0 width.

12.1 Grand unification

The Q^2 ($\equiv |q^2|$) dependence of electroweak and strong coupling constants implies that they may converge at very high energies. For example, the fine-structure constant α is 1/137 when measured at low Q^2 and increases with Q^2: α_s on the other hand is of order unity at low Q^2 and falls with increasing Q^2. If the coupling constants do converge this would mark a further stage of the unification process, known as grand unification (see Ramond (1983)). What could the unifying group (G) be, and at what energy does grand unification occur? It is necessary to be able to embed $SU(3)_c \otimes SU(2)_L \otimes U(1)$ in G, hence G should possess at least four commuting generators to match t_3, y, I_3^c and Y^c, i.e. it must have at least rank 4. The simplest example for G with rank 4 is SU(5); if the simple choice for the Higgs bosons of the standard model is retained we have minimal SU(5). There are several candidate groups for G with higher rank which contain SU(5). SU(5) serves to illustrate several points likely to be true, however, for any successful grand-unification group. First, we establish the scale of grand unification. The couplings strengths α_i for SU(i) gauge couplings can be shown to vary with Q^2 in the following way

$$\alpha_i^{-1}(Q)^2 = \alpha_i^{-1}(\mu^2) + \ln(Q^2/\mu^2)[11n_b - 4n_f]/12\pi \qquad [12.1]$$

where $(11n_b - 4n_f)$ indicates the contribution of boson (fermion) loops to vacuum polarization. This combines the expressions relevant to QCD and to the electroweak interactions. n_b is 0, 2 and 3 for U(1), SU(2) and SU(3) respectively; while n_f, the number of active fermion generations, is taken to be 3. At the scale Q^2 which gives unification

$$(5/3)\alpha_1(Q^2) = \alpha_2(Q^2) = \alpha_3(Q^2),$$

which leads to two predictions:

$$Q\ \text{(unification)} \simeq 10^{15}\ \text{GeV}$$

$$\sin^2\theta_W(M_W^2) \simeq 0{\cdot}214,$$

so that $\sin^2\theta_W$ is now predicted correctly. Figure 12.1 shows the variation of $5\alpha_1/3$, α_2 and α_3 extrapolated with energy according to equation [12.1]. At energies above 10^{15} GeV all the interactions are unified into a semi-simple compact group G which breaks spontaneously down into $SU(3)_c \otimes SU(2)_L \otimes U(1)_y$ at this energy by the Higgs mechanism. At 100 GeV energy the breakdown of the standard model by a second Higgs mechanism occurs. Thus schematically:

$$G \rightarrow SU(3)_c \otimes SU(2)_L \otimes U(1)_y$$
$$\rightarrow SU(3)_c \otimes U(1)_{em}.$$

The left-handed fermions and their antifermions from one

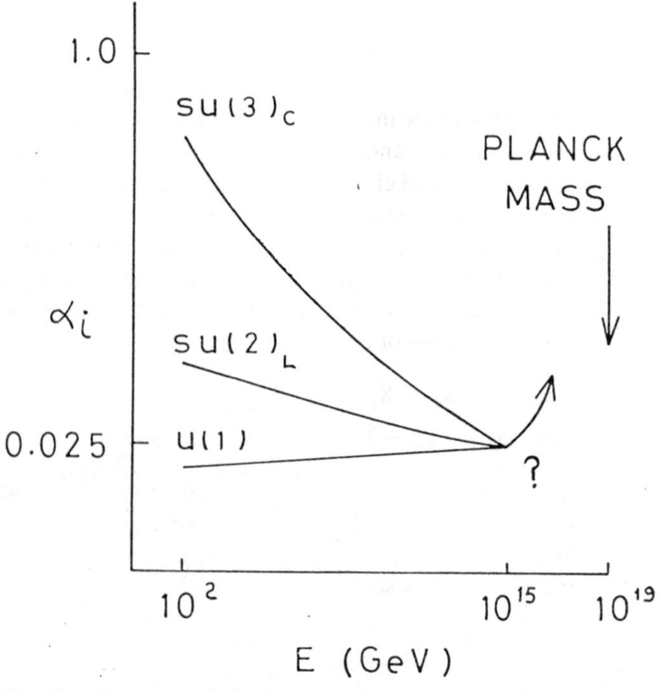

Fig. 12.1 Variation of the coupling strengths $5\alpha_1/3$, α_2 and α_3 with energy in a grand-unification scheme.

generation occupy one $\bar{5}$ and one 10-multiplet of SU(5):

$$\bar{5}\ (\bar{\mathrm{d}}_{\mathrm{b}}, \bar{\mathrm{d}}_{\mathrm{y}}, \bar{\mathrm{d}}_{\mathrm{r}}, \mathrm{e}^-, \nu_{\mathrm{e}})$$
$$10\ (\bar{\mathrm{u}}_{\mathrm{b}}, \bar{\mathrm{u}}_{\mathrm{y}}, \bar{\mathrm{u}}_{\mathrm{r}}, \mathrm{d}_{\mathrm{b}}, \mathrm{d}_{\mathrm{y}}, \mathrm{d}_{\mathrm{r}}, \mathrm{u}_{\mathrm{b}}, \mathrm{u}_{\mathrm{y}}, \mathrm{u}_{\mathrm{r}}, \mathrm{e}^+).$$

This certainly unites a generation, but with the unexpected feature of putting quarks and antiquarks in the same multiplet.

It is now possible to explain the exact equality of the proton and electron charges. In an Abelian gauge theory (U(1)) the eigenvalue of the generator (charge) is a continuous variable and may in principle have any value. On the other hand the generators of the non-Abelian SU(5) groups only have discrete eigenvalues: charge is included amongst them and is therefore quantized. The sum of eigenvalues of the charge generator is zero for each representation, so that in the case of the $\bar{5}$:

$$-3q_{\mathrm{d}}e - e = 0$$
$$q_{\mathrm{d}} = -1/3.$$

This provides a new insight: there is a connection between the number of colours and the fractional charge of quarks. The exact cancellation of electron and proton charges then follows. There are twenty-four gauge bosons in SU(5) ((5^2-1) being the dimension of the self-adjoint representation). Of these twelve are familiar: the eight gluons, $\mathrm{W}^{\pm}$, Z^0 and photon of the standard model. The twelve new gauge bosons are the X and Y bosons with masses of order 10^{15} GeV/c^2:

	X_i	$\bar{\mathrm{X}}_i$	Y_i	$\bar{\mathrm{Y}}_i$
charge	$+\frac{4}{3}$	$-\frac{4}{3}$	$+\frac{1}{3}$	$-\frac{1}{3}$

where i is the colour index. Typical two-body processes involving these so called leptoquarks are shown in Fig. 12.2. Quark conservation is violated and the proton, which we have hitherto regarded as stable, would decay via:

$$\mathrm{p} \to \pi^+ + \bar{\nu}_{\mathrm{e}}$$
$$\mathrm{u(ud)} \to \mathrm{u}(\bar{\nu}_{\mathrm{e}}\bar{\mathrm{d}}) \quad \text{(at the quark level)}$$

or

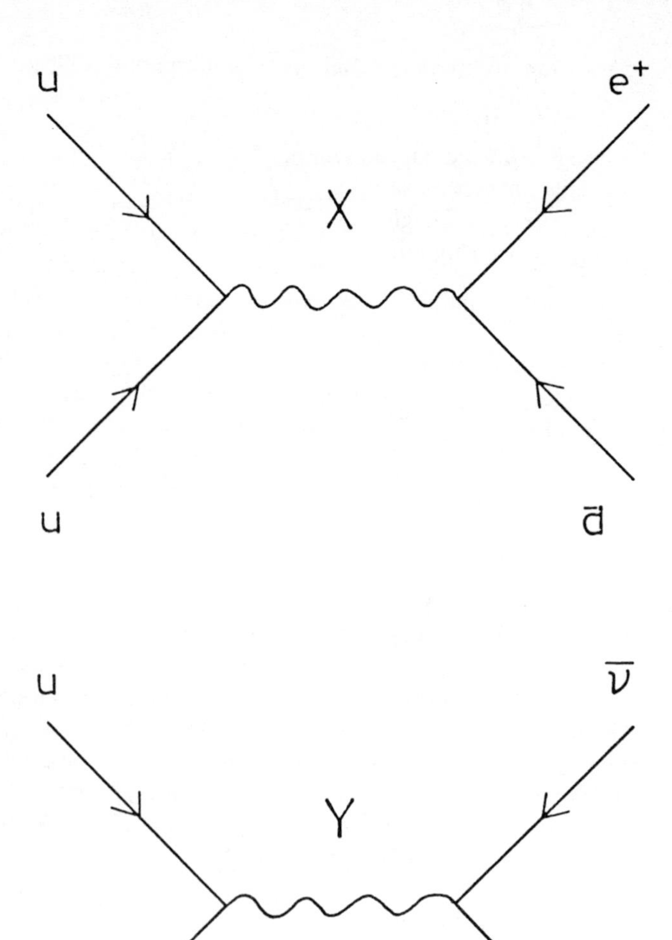

Fig. 12.2 Quark-quark interactions involving the heavy gauge bosons X and Y (leptoquarks).

$$p \rightarrow e^+ + \pi^0$$
$$(uu)d \rightarrow (e^+\bar{d})d \quad \text{(at the quark level).}$$

Although allowing baryon number (B) and lepton number (L) to change, processes satisfying the SU(5) symmetry conserve the difference $B-L$ globally. The calculated lifetime of the proton is extremely long

$$\tau_p \simeq m_x^4/\alpha^2 m_p^5 \sim 10^{30} \text{ years,}$$

as it must be in order that proton decay might have avoided notice until the present. From minimal SU(5) the rate for the mode $p \rightarrow e^+ + \pi^0$ is expected to be greater than 10^{-31}/yr (Perkins (1984)). In other words each year in 16 tonnes of hydrogen one proton at least should on average decay to $e^+\pi^0$. One technique for detecting such decays is to view a huge bath of pure water with photomultipliers which would register the Cerenkov radiation from the decay positrons or muons. The apparatus needs to be located deep underground in order to reduce cosmic ray background. Several hundred tonne-years of observation have been accumulated without any examples of $e^+\pi^0$ decay being observed. This fixes the rate to be less than 5×10^{-33}/year (Vergados (1986)), so the simplest SU(5) grand-unification scheme is ruled out. Other candidate schemes based on unification groups of higher rank are not excluded. More sophisticated and larger experiments should push the observational limit to 10^{-33}/year for the $e^+\pi^0$ mode. At this level the background from $\bar{\nu}_e + p \rightarrow e^+ + \pi^0 + n$ initiated by cosmic ray antineutrinos overwhelms the signal. Whether the proton decays, remains an open question; its answer is crucial to the future of grand unified theories. In the event, grand unified theories have other serious problems.

The two unification scales (M_W) and (M_X) are so enormously different, by a factor 10^{13}, that it is very hard to ensure that the scales remain isolated: for example, radiative corrections of scale M_X affect the mass of the electroweak Higgs boson. This is known as the 'hierarchy problem'. It would appear to require a delicate and unlikely set of cancellations in field theory. Supersymmetric theories seem to answer this difficulty by providing means to isolate the two scales; they do not explain the disparity between the scales.

12.2 Supersymmetry

Supersymmetry provides a way of connecting bosons and fermions (Wess and Zumino (1974)). The standard generators of the Poincaré group generate translations, rotations and boosts. To these are added N spinorial charges (Q_i) which can convert fermions into bosons and vice versa. This is the *only* way that an internal symmetry and the Poincaré group can be embedded in a higher symmetry. The theory with N equal to unity is termed simple supersymmetry and theories with larger values of N are termed extended supersymmetries.

The basic supermultiplets for simple supersymmetry have a 'gauge' multiplet with spin content $(1, \frac{1}{2})$ and a 'chiral' multiplet with spin content $(\frac{1}{2}, 0)$. Each particle possesses a new quantum number called R parity. R is zero for all the currently known particles and would be ± 1 for their supersymmetric partners. R is conserved in all interactions, hence the supersymmetric partners would have to be produced in pairs in any interaction between particles having R equal to zero. Table 12.2 lists the supersymmetric doublets for simple supersymmetry.

In the simplest view the particle and 'sparticle' should have equal masses. Cancellation of radiative corrections due to loops is a characteristic feature of supersymmetric theories. Boson and fermion loops contribute with opposite sign, therefore if every fermion is partnered by a boson and vice versa there is exact cancellation. The result is that it becomes possible to isolate the unification scales. In addition the increase in the number of fermions in supersymmetric theories leads to a slower energy dependence of the coupling strengths (equation [12.1]). The grand-unification scale moves to 10^{17} GeV and this entrains an increase in the proton lifetime. It should be possible to produce sparticles in e^+e^- collisions via:

$$
\begin{aligned}
e^+ + e^- &\to \tilde{l} + \tilde{\bar{l}} \\
&\qquad\quad \hookrightarrow \bar{l} + \tilde{\gamma} \\
&\quad \hookrightarrow l + \tilde{\gamma}
\end{aligned}
$$

Table 12.2

Spin 1	*Spin* $\frac{1}{2}$		*Spin* 0	
Gluon	Gluino	($\tilde{g}$)		
Photon	Photino	($\tilde{\gamma}$)		
$W^{\pm}$	Wino	($\tilde{W}$)		
Z^0	Zino	($\tilde{Z}$)		
	Higgsino	($\tilde{H}$)	Higgs	(ϕ_0)
	Lepton		slepton	($\tilde{l}$)
	Quark		squark	($\tilde{q}$)

Spin 2	*Spin* $\frac{3}{2}$
graviton	gravitino

or in $\bar{p}p$ collisions via:

$$p+\bar{p}\rightarrow\tilde{q}+\tilde{\bar{q}}+\text{anything}$$

$$\tilde{\bar{q}}\rightarrow\bar{q}+\tilde{g}$$

$$\tilde{q}\rightarrow q+\tilde{g}.$$

Sparticles once produced then decay to other sparticles and eventually to the lightest, presumably the photino, whose interactions are comparable to those of a neutrino. Consequently the signature for sparticle production is a substantial 'missing energy' carried off by photinos. Searches at current accelerator energies have not found any positive evidence for sparticle production, and this implies that supersymmetric particles may be very massive (e.g. $M(\tilde{\gamma})>$ 20 GeV/c^2 and $M(\tilde{g})$, $M(\tilde{q})>40$ GeV/c^2).

12.3 Cosmology

The accepted model is of a universe expanding continuously from an initial space-time point. Observation shows that the recession velocity of galaxies is proportional to their distance from us

$$\mathrm{d}R/\mathrm{d}t=RH(t)$$

where $H(t)$ is called Hubble's constant and has a value 75 ± 25 km s^{-1} Mpc^{-1} at the present epoch. (1 pc is 3·26 light years.) The age of the universe is of order H^{-1}, i.e. 1.3×10^{10} years. Inside the solar system there is marked asymmetry with matter being present but little or no antimatter. Beyond the solar system there is no evidence of annihilation processes between stars or galaxies and anti-stars or galaxies; nor do antiprotons appear to be present at a significant level among primary cosmic rays. The asymmetry, therefore, seems to persist through the universe. There are 10^9 as many photons as protons, forming a cosmic microwave background with a temperature of 2·7 K. This radiation is a relic from the early universe when most energy was in the form of radiation. Today, 10^{10} years later, the radiation has cooled and its energy is only 10^{-4} of the energy in matter. Matter is mainly in the form of hydrogen with a substantial helium component (25 per cent relative abundance). Nucleosynthesis in stars is incapable of accounting for this amount of helium, which must have been formed in the early moments of the universe.

From these selected features of the universe a number of inferences important to particle physics can be drawn. The current asymmetry between matter and antimatter requires three conditions to hold: there must be baryon-number violating reactions, for which CP and C must both be violated and finally there must have been a departure from thermal equilibrium in the universe (Sakharov (1967)). An example will show why all these conditions are necessary. Consider the case where baryon violation occurs in X-decay, for which two channels are available

$$\begin{aligned} \mathrm{X} &\rightarrow \mathrm{B}_1 \quad \text{branching fraction } r \\ &\rightarrow \mathrm{B}_2 \quad \text{branching fraction } (1-r); \end{aligned}$$

while for the antiparticle

$$\begin{aligned} \bar{\mathrm{X}} &\rightarrow -\mathrm{B}_1 \quad \text{branching fraction } \bar{r} \\ \bar{\mathrm{X}} &\rightarrow -\mathrm{B}_2 \quad \text{branching fraction } (1-\bar{r}). \end{aligned}$$

B_1, B_2, $-\mathrm{B}_1$ and $-\mathrm{B}_2$ specify the number of baryons in the final states. When equal numbers of X and $\bar{X}$ decay the baryon-number asymmetry generated is

$$\varepsilon = (r - \bar{r})(B_1 - B_2),$$

which will be non-zero only if $B_1 \neq B_2$ and $\bar{r} \neq r$. The latter condition requires *C* violation at least. However, if *C* is violated but not *CP* then only the angular distributions of $\bar{X}$ and X-decay differ and not their rates. When the temperature of the universe is high (10^{15} GeV) the parents and their decay products remain in thermal equilibrium and the corresponding asymmetries in the reverse reactions ensure that all the baryon asymmetry is washed out. It was therefore not enough for the temperature of the universe to fall below 10^{15} GeV: it must have been falling rapidly compared to the rate (Γ_x) of the processes $X \rightleftharpoons B$. In the case that the temperature declined rapidly the decays of X continued to occur but the average energy of the baryons would have become insufficient to produce any X-leptoquarks; thus a baryon asymmetry would be 'frozen in'.

A similar departure from thermal equilibrium must have occurred later in the history of the universe when the temperature, *T*, fell below that necessary to maintain thermal equilibrium between neutrons and protons via reactions such as

$$e^- p \rightleftharpoons n\nu_e, \quad e^+ n \rightleftharpoons p\bar{\nu}_e, \quad \text{and} \quad n \to pe^- \bar{\nu}_e.$$

In that era (t) the characteristic time ($t_w \simeq G^{-2}T^{-5}$) for weak interactions became large compared to $H^{-1}(t)$. The primordial neutrons that remained were soon (relative to the neutron lifetime) trapped into helium nuclei. An important corollary is that the larger the number of particle species existing in thermal equilibrium the higher the expansion rate of the universe and the earlier the neutron ratio is frozen in. It follows that the number of neutrino species has a direct influence on the neutron abundance and hence on the helium abundance. It has been calculated that each additional neutrino species leads to an increase in the helium abundance by about 1·5 per cent. From the measured helium abundance the astrophysicists are able to put a tentative limit of 3 or 4 to the number of light neutrino species. In the context of the standard model this implies that there might be one further generation of leptons and quarks to uncover.

A more controversial subject is that of the neutrino mass. At present the matter in the universe which is emitting radiation seems to account for less than 10 per cent of the total gravitational mass in the universe. In particular the motion of gas clouds around a typical galaxy under its gravitational attraction demands much more mass in the galaxy than is visible. The large amount of 'dark matter' could be in the form of neutrinos if these were to possess mass. Making the assumption that all the dark matter is accounted for by massive neutrinos leads to a useful upper bound on this mass; if a sum is made over the neutrino species, then

$$\Sigma m(\nu) < 100 \text{ eV}/c^2.$$

The possibility of neutrinos having non-zero mass has several important consequences, which are being tested in terrestrial experiments. In principle the simplest experiment is to measure precisely the end point of the spectrum of the electrons emitted in ^{3}H β-decay; this end point is only 18·6 KeV. Should the neutrino have mass, m_ν, then the maximum electron kinematic energy would be reduced by m_ν. At the moment there is disagreement about the effects on this result of experimental resolution and of using ^{3}H bound in molecules. It can be fairly stated only that

$$m(\nu_e) < 10 \text{ eV}/c^2.$$

Another consequence of having massive neutrinos would be the possibility that the lepton mass eigenstates could differ from the weak eigenstates, just as is the case for the quarks. The weak eigenstates could be expressed as linear superpositions of mass eigenstates:

$$\nu_e = \nu_1 \cos\theta + \nu_2 \sin\theta$$
$$\nu_\mu = \nu_2 \cos\theta - \nu_1 \sin\theta,$$

where for simplicity we consider only two lepton generations. A pure $\bar{\nu}_e$ beam from a reactor would have, on emergence at time zero,

$$\bar{\nu}_e(0) = \bar{\nu}_1(0) \cos\theta + \bar{\nu}_2(0) \sin\theta$$
$$0 = \bar{\nu}_2(0) \cos\theta - \bar{\nu}_1(0) \sin\theta.$$

Thus

$$\bar{\nu}_e(0) = \bar{\nu}_1(0)/\cos\theta = \bar{\nu}_2(0)/\sin\theta.$$

Each eigenstate develops in space-time like:

$$\nu_i(t, x) = \nu_i(0, 0) \exp i[-E_i t + p_i x].$$

The momenta must be identical in order for the eigenstates to be coherent at time zero, but the energies can differ. Then, after time t,

$$\begin{aligned}\bar{\nu}_e(t) &= \bar{\nu}_1(t) \cos\theta + \bar{\nu}_2(t) \sin\theta \\ &= \bar{\nu}_1(0) \cos\theta \exp(-iE_1 t) + \bar{\nu}_2(0) \sin\theta \exp(-iE_2 t) \\ &= \bar{\nu}_e(0) [\cos^2\theta \exp(-iE_1 t) + \sin^2\theta \exp(-iE_2 t)],\end{aligned}$$

where the overall factor $\exp(ipx)$ has been eliminated. The relative intensity at time t is therefore

$$\begin{aligned}I(t) &= \cos^4\theta + \sin^4\theta \\ &\quad + \sin^2\theta \cos^2\theta [\exp(i\Delta E t) + \exp(-i\Delta E t)] \\ &= 1 - \sin^2 2\theta \sin^2[\Delta E t/2],\end{aligned}$$

where $\Delta E = E_1 - E_2$. The neutrino mass is expected to be much smaller than the energy (of order 1 MeV), which means that E can be approximated by $p + m^2/2p$. Writing Δm^2 for $m_1^2 - m_2^2$ gives

$$I(t) = 1 - \sin^2 2\theta \sin^2[\Delta m^2 t/4p].$$

The intensity of each weak species is oscillating as the beam travels away from the source, with a periodicity:

$$\begin{aligned}l &= 4\pi p/\Delta m^2 \\ &= 2{\cdot}476[p(\text{MeV}/c)/\Delta m^2((\text{eV}/c^2)^2)] \text{ metres.}\end{aligned}$$

This gives an oscillation length of order 100 metres if Δm^2 is around 100 $(\text{eV}/c^2)^2$. In one experiment the antineutrinos are detected using a combination of liquid scintillator and wire chambers filled with ^{3}He. The positrons from $\bar{\nu}\text{p} \rightarrow \text{ne}^+$ are detected in the scintillator; and the neutrons from this reaction are detected in the ^{3}He chambers indirectly through the observation of recoils from n-^{3}He collisions. Oscillations may be searched for either by comparing the rate of electron-antineutrino interactions with that expected for the known reactor flux or by making measurements with the detector at two different distances from the reactor. There is as yet no evidence for neutrino oscillations.

Finally if neutrinos were to possess mass it would be likely that they would be Majorana particles. Majorana neutrinos have the property of being identical to their antiparticles (Frampton (1982)). Then neutrinoless double β-decay is possible but is suppressed by the fact that the weak interaction has a $V-A$ character. Lifetimes are extremely long ($\sim 10^{20}$ years), which makes experiments very difficult. The existence of neutrinoless double β-decay would be firmly established by the observation of decays in which the sum of the energy of the two β's equalled the mass difference between parent and daughter nucleus. At present very little is known about double β-decay and there is no evidence in favour of neutrinoless double β-decay (Vergados (1986)). To conclude, it is of prime importance for experiments to continue to make more refined tests on the nature of neutrinos. In some grand-unification schemes a neutrino mass appears in a natural way.

12.4 Gravitation

After grand unification the next logical step would be to seek unification of gravity with the other forces. Classical general relativity is already a gauge theory (Misner (1970)). The relevant scale for unification may be that at which the gravitational self-energy of a mass M becomes equal to its rest-mass energy:

$$G_N M^2/\lambda \approx M,$$

where G_N is the gravitational constant and λ, the Compton wavelength, is equal to M^{-1}. The corresponding energy is called the Planck energy:

$$M_{\mathrm{p}} = G_{\mathrm{N}}^{-\frac{1}{2}} = 10^{19}\ \mathrm{GeV}.$$

A related point to note is that in supersymmetric theories the scale at which grand unification occurs is of order 10^{17} GeV.

In supersymmetric algebras the anti-commutator for the spinorial charges is $\{Q_i, Q_j\} = \sigma_\mu p_\mu \delta_{ij}$, where p_μ, the momentum operator, is the generator for translations. This provides a link to general relativity: invariance under translations is an important ingredient in classical general relativity. When super symmetries are treated as local rather than as global and are gauged, a new spin-2 gauge field and its supersymmetric spin-$\frac{3}{2}$ partner emerge. The former can be identified with the gravitational field and its quantum is the spin-2 graviton (see Table 12.2). The $N=1$ supersymmetric theory has been christened supergravity.

All *local* quantum field theories of gravity are doomed because of quantum fluctuations, which have a range equal to the Planck length (M_{P}^{-1}) of $1{\cdot}6 \times 10^{-35}$ m. These quantum fluctuations make it impossible to construct a renormalizable field theory along the usual lines. Fortunately another class of theory has come into play (Schwarz (1984, 1985) using finite one-dimensional strings rather than points to describe particles. In this way the difficulty of localization is avoided. Supersymmetric string theories turn out to be renormalizable in ten dimensions! It is necessary to picture the excess six dimensions to be compact, i.e. forming closed loops of size 10^{-35} m or less. Although strings are non-local their interactions are local; interactions take place at points in space-time. It is claimed that superstring theory may lead to a complete theory of the elementary particles and elementary forces; it will be interesting to watch developments.

Note added in proof

The observation of neutrino induced reactions in coincidence with the supernova 1987a has had an impact on our understanding of neutrinos. The source is 50 kpc from earth. Some dozen interactions were observed in a burst lasting a few seconds at apparatus being used to search for proton decays. Over the flight path to earth neutrinos with mass would travel slower than light. For a neutrino of mass m_ν eV/c^2 and energy E MeV the delay would be 2·5 $(m_\nu/E)^2$ seconds. Thus earlier interactions would be more energetic. No obvious correlation between energy release and time was seen: it has been inferred that the neutrino mass is less than 10 eV/c^2. It is also clear that to survive the journey the neutrino lifetime must be very long—if it is not stable.

Appendix A
Two-body kinematics, cross-sections and decay rates

We consider the two-body process

$$a+b \rightarrow c+d$$

and let $(E, \mathbf{p})$ be the notation for four-momentum. The total CM energy squared

$$W^2 \equiv s=(E_a+E_b)^2-(\mathbf{p}_a+\mathbf{p}_b)^2.$$

If b is at rest (the target)

$$s=m_a^2+2E_a m_b+m_b^2,$$

so that if all masses are small compared to E_a,

$$s \simeq 2E_a m_b.$$

(Note that m_b is the target mass.) The boost from the laboratory to the CM frame has velocity

$$\bar{\beta}=p_a/(E_a+m_b) \quad \text{and} \quad \bar{\gamma}=(E_a+m_b)/W.$$

If, on the other hand, the collisions take place between identical particles with equal and opposite three-momenta then

$$s=4E^2,$$

where $E_a=E_b=E$. The laboratory and CM frames coincide in this case.

The three-momenta of c and d are equal and opposite in the CM rest frame and have magnitude

$$p_c^*=p_d^*=\{[W^2-(m_c+m_d)^2][W^2-(m_c-m_d)^2]\}^{1/2}/2W$$

and

$$E^*_{c,d} = (W^2 \pm m_c^2 \pm m_d^2)/2W,$$

where an asterisk is used to denote CM quantities. The four-momentum transfer is defined as

$$\begin{aligned} q = (q_0, \mathbf{q}) &= (E_a - E_c, \mathbf{p}_a - \mathbf{p}_c) \\ &= (E_d - E_b, \mathbf{p}_d - \mathbf{p}_b). \end{aligned}$$

The covariant quantity q^2 is usually less than or equal to zero. In the case of elastic scattering through an angle θ^* in the CM frame

$$\begin{aligned} q^2 &= 0^2 - (\mathbf{p}^*_a - \mathbf{p}^*_c)^2 \\ &= -2p^{*2}(1 - \cos\theta^*) \end{aligned}$$

because $p^*_a = p^*_c = p^*$ for elastic scattering. Then q^2 is exactly zero in the forward direction. Other notations employ the symbols Q^2 and t, where

$$Q^2 = -t = -q^2.$$

In many of the processes considered in the text, the momenta of the particles are so large that it is reasonable to neglect the masses. Then all the CM energies and momenta have equal magnitude:

$$p^*_a = p^*_b = p^*_c = p^*_d = p^*,$$
$$E^*_a = E^*_b = E^*_c = E^*_d = E^*,$$
$$E^* = p^* = W/2 = \sqrt{s}/2 \text{ and}$$
$$q^2 = -2p^{*2}(1 - \cos\theta^*) = -(s/2)(1 - \cos\theta^*).$$

Most experiments in particle physics make determinations of reaction cross-sections or particle lifetimes. Suppose a particular experiment involves the detection of mesons emitted from a hydrogen target bombarded by a beam of particles. The number of mesons detected (ΔN) will be strictly proportional to:

(1) the time duration of the experiment (τ),
(2) the volume of target illuminated by the beam (V),

(3) the number of beam particles incident per second (I),
(4) the solid angle subtended by the detector at the target ($\Delta\Omega$), and
(5) the density of protons in the target (ρ).

Thus

$$\Delta N \propto (I\rho)(V\tau)\Delta\Omega.$$

The constant of proportionality has dimensions of (area/solid angle) and is the effective scattering area presented by one target proton for meson emission in the direction of the detector. This is known as the differential cross-section ($\mathrm{d}\sigma/\mathrm{d}\Omega$) and is measured in m^2/sterad or barns/sterad (1 barn is $10^{-24}\ \mathrm{m}^2$). Then

$$\Delta N = (\mathrm{d}\sigma/\mathrm{d}\Omega) I\rho V\tau\Delta\Omega.$$

Integration over the whole solid angle yields the total cross-section for emission. If a detector is placed behind the target the number of beam particles removed by all interactions can be measured. This determines the total cross-section, σ_{T}. Some interactions lead to a final state in which there are only two particles and these are identical to the beam and target. These are elastic collisions with a total cross-section σ_{EL}. The remaining cross-section is the inelastic or reaction cross-section, σ_{R} involving processes in which the outgoing particles are not simply the beams plus target. Then

$$\sigma_{\mathrm{T}} = \sigma_{\mathrm{EL}} + \sigma_{\mathrm{R}}.$$

Cross-sections may be calculated from the Feynman graph amplitudes as follows. $|T|^2$ is the transition probability for a single beam and target to interact and produce a particular final state. Let us suppose there are n particles in the final state. Then, following the lines of the argument made above,

$$|T|^2 = (\mathrm{d}\sigma/\mathrm{d}Q_n)(T\rho V\tau),$$

where Q_n is the density of distinguishable quantum mechanical states having the configuration of the n outgoing particles. With the normalization adopted in Chapter 5 for the plane-wave particle states, $V\tau$ reduces to unity and $I\rho$ reduces to a flux factor F. We have

$$d\sigma = |T|^2 \, dQ_n/F,$$

where dQ_n is called the element of n-body phase space. All three quantities $|T|^2$, dQ_n and F are invariant under Lorentz transformations. In the *CM frame*

$$F = 4q_i^* W,$$

where q_i^* is the magnitude of the three-momentum of either incident particle. Evaluating dQ_2 in the *CM frame* for a two-body final state gives

$$dQ_2 = q_f^* d\Omega^*/16\pi^2 W,$$

where q_f^* is the magnitude of the three-momentum for an outgoing particle pointing into the solid angle $d\Omega^*$. Then

$$d\sigma = |T|^2 q_f^* \, d\Omega^*/64\pi^2 q_i^* W^2.$$

If the momenta of the particles are so large that it is reasonable to neglect masses, the above results become:

$$F = 2s,$$
$$dQ_2 = d\Omega^*/32\pi^2,$$
$$d\sigma = |T|^2 \, d\Omega^*/64\pi^2 s.$$

Decay rates of particles are given in terms of the matrix elements, T, by similar expressions with the flux factor replaced by $2M$, where M is the parent mass. The decay rate is thus

$$d\Gamma = |T|^2 \, dQ_n/2M,$$

which for a two-body decay becomes

$$d\Gamma = |T|^2 q_f^* d\Omega^*/32\pi^2 M^2.$$

If the initial- and final-state particles have spin then T may depend on the polarization state of each of them. In the case of a reaction $1+2 \rightarrow 3+4$, T becomes $T(m_1, m_2, m_3, m_4)$, where m_1, m_2, m_3 and m_4 are the spin components along some quantization axis. Measurements of cross-section are often made with an unpolarized beam and target and with no attempt to identify the final-state polarizations. In this simple case the spin average must be taken for the initial state and the spin sum for the final state. Then

$$\mathrm{d}\sigma = \frac{1}{N_\mathrm{f}} \sum_{\mathrm{m}} |T(m_1, m_2, m_3, m_4)|^2 (q_\mathrm{f}^* \mathrm{d}\Omega^* / 64\pi^2 q_\mathrm{i}^* W^2),$$

where the spin factor $N_\mathrm{f} = (2s_1 + 1)(2s_2 + 1)$ if the incident particles have spin s_1 and s_2. More details on phase space and cross-section calculations can be found in A. D. Martin and T. D. Spearman, *Elementary Particle Theory*, North-Holland, Amsterdam, 1970.

Appendix B
The Breit-Wigner formula

The wave-function of an unstable particle (resonance) in its own rest frame is

$$\psi(t)=\psi(0)\exp[-iM_0t-\Gamma t/2].$$

Γ determines the probability $P(t)$ for survival beyond a time t:

$$P(t)=|\psi(t)|^2/|\psi(0)|^2=\exp(-\Gamma t).$$

Thus the decay rate is Γ. M_0 is a frequency and, because $\hbar$ and c are set to unity, it is equally the particle mass. The energy wavefunction is the Fourier transform of $\psi(t)$:

$$\begin{aligned}\phi(M)&=\int_0^\infty \mathrm{d}t\, \psi(t)\exp(iMt)\\ &=i/(M_0-M-i\Gamma/2).\end{aligned}$$

Reference to section 4.2 shows that $\phi(M)$ has precisely the form expected for the propagator of a particle having a complex mass equal to $(M_0-i\Gamma/2)$. The probability of the resonance having an observed mass in the range from M to $M+\mathrm{d}M$ is

$$P(M)\,\mathrm{d}M\propto|\phi(M)|^2\mathrm{d}M,$$

so that

$$P(M)\propto 1/[(M_0-M)^2+\Gamma^2/4].$$

This last expression is the non-relativistic Breit-Wigner resonance formula. When Γ is small compared to M_0 the mass distribution is a sharp peak centred at M_0 and the full width of the distribution at half maximum is Γ. Thus the more rapid the decay, the wider the resonance. Classically Γ is the damping coefficient. Each decay channel (i) makes a contribution to the decay rate (the partial width, Γ_i) and

$$\Gamma = \sum_i \Gamma_i.$$

Appendix C
Clebsch-Gordan coefficients for SU(2)

Clebsch-Gordan coefficients for SU(2) or 0(3), with the usual sign convention are given below. Further tables can be found in Aguilar-Benitez *et al.*, *Rev. Mod. Phys.* **56** (April 1984). The notation is $(J, M|j_1, m_1; j_2, m_2)$.

$j_1 = j_2 = \frac{1}{2}$: $J = 0$ *or* 1

$J=$	1		
$M=$	± 1	m_1	m_2
	1	$\pm\frac{1}{2}$	$\pm\frac{1}{2}$

$J=$	1	0		
$M=$	0	0	m_1	m_2
	$\sqrt{\frac{1}{2}}$	$\sqrt{\frac{1}{2}}$	$+\frac{1}{2}$	$-\frac{1}{2}$
	$\sqrt{\frac{1}{2}}$	$-\sqrt{\frac{1}{2}}$	$-\frac{1}{2}$	$+\frac{1}{2}$

$j_1 = 1, j_2 = \frac{1}{2}$: $J = \frac{3}{2}$ or $\frac{1}{2}$

$J=$	$\frac{3}{2}$		
$M=$	$\pm\frac{3}{2}$	m_1	m_2
	1	± 1	$\pm\frac{1}{2}$

$J=$	$\frac{3}{2}$	$\frac{1}{2}$		
$M=$	$\frac{1}{2}$	m_1	m_2	
	$\sqrt{\frac{1}{3}}$	$\sqrt{\frac{2}{3}}$	1	$-\frac{1}{2}$
	$\sqrt{\frac{2}{3}}$	$-\sqrt{\frac{1}{3}}$	0	$+\frac{1}{2}$

$J=$	$\frac{3}{2}$	$\frac{1}{2}$		
$M=$	$-\frac{1}{2}$	$-\frac{1}{2}$	m_1	m_2
	$\sqrt{\frac{2}{3}}$	$\sqrt{\frac{1}{2}}$	0	$-\frac{1}{2}$
	$\sqrt{\frac{1}{3}}$	$\sqrt{\frac{2}{3}}$	-1	$+\frac{1}{2}$

Appendix D
The Dirac equation and its solutions

In non-relativistic quantum mechanics the reader may recall that Schrödinger's equation was obtained by converting the energy equation into operator form. Thus

$$E = p^2/2m + U(\text{potential energy})$$

with the operator substitutions $E \to i\partial/\partial t$ and $p \to -i\nabla$ becomes:

$$i\partial/\partial t = -\nabla^2/2m + U,$$

which operates on the wavefunction ψ. The corresponding relativistic energy equation for a free particle is $E^2 - p^2 - m^2 = 0$, which converts to the Klein-Gordan equation

$$(-\partial^2/\partial t^2 + \nabla^2 - m^2)\psi = 0.$$

The simplest solutions of this equation describe massive bosons with spin zero. However, Dirac discovered a *linear* relativistic wave equation whose solutions describe free particles with spin $\frac{1}{2}$ and Landé g-factor equal to 2. This is the appropriate equation for free electrons. In standard notation the Dirac equation is:

$$(i\gamma_\mu \partial_\mu - m)\psi = 0,$$

where γ_μ is as yet some undetermined four-vector-like quantity and ψ the wavefunction. ∂_μ is shorthand for the four-vector $\partial/\partial x_\mu$ with components $\partial/\partial t$, $-\partial/\partial x$, $-\partial/\partial y$, $-\partial/\partial z$. The product $\gamma_\mu \partial_\mu$ is the scalar product of special relativity

$$\begin{aligned}\gamma_\mu \partial_\mu &= \gamma_0 \partial/\partial t + \gamma_1 \partial/\partial x + \gamma_2 \partial/\partial y + \gamma_3 \partial/\partial z \\ &= \gamma_0 \partial/\partial t + \boldsymbol{\gamma} \cdot \boldsymbol{\nabla}.\end{aligned}$$

An alternative is to write $g_{\mu\nu}\gamma_\mu\partial_\nu$ for this product, where $g_{\mu\nu}$ has the components

$$g_{\mu\nu} = \begin{bmatrix} 1 & \cdot & \cdot & \cdot \\ \cdot & -1 & \cdot & \cdot \\ \cdot & \cdot & -1 & \cdot \\ \cdot & \cdot & \cdot & -1 \end{bmatrix}$$

with all the off-diagonal elements being zero. Applying $(i\gamma_\mu\partial_\mu + m)$ to the Dirac equation gives

$$(i\gamma_\mu\partial_\mu + m)(i\gamma_\rho\partial_\rho - m)\psi = 0$$

i.e.

$$(-\gamma_\mu\gamma_\rho\partial_\mu\partial_\rho - m^2)\psi = 0,$$

which will have to reduce to the Klein-Gordan equation because electrons satisfy the relativistic energy equation. This reduction can only take place if

$$\gamma_0^2 = 1, \quad \gamma_i^2 = -1 \quad (i = 1, 2, 3)$$

and

$$\gamma_\mu\gamma_\rho + \gamma_\rho\gamma_\mu = 0 \quad (\mu \neq \rho).$$

The surprising result here is that the components of γ do not commute but *anti-commute* and must, therefore, be matrices. Summarizing:

$$\gamma_\mu\gamma_\rho + \gamma_\rho\gamma_\mu = \{\gamma_\mu, \gamma_\rho\} = 2g_{\mu\rho}.$$

ψ factorizes into a piece describing the position of the electron in space-time and a piece (u) describing whether the electron spin is up or down:

$$\psi = u \exp[-i(Et - \mathbf{p} \cdot \mathbf{x})] = u \exp(-\mathrm{i}p_\rho x_\rho),$$

where the four-momentum p_ρ has components $(E, \mathbf{p})$. The choice of a plane wave is justified here by the fact that any wavefunction can be Fourier analysed into a superposition of plane waves: hence results proved for a plane wave are generally true. With this choice Dirac's equation yields

$$(i\gamma_0\partial/\partial t + i\gamma_i\partial/\partial x_i - m)u \exp(-iEt + ip_i x_i) = 0$$

$$\therefore \qquad (\gamma_0 E - \gamma_i p_i - m)u = 0$$

$$\text{i.e.} \qquad (\gamma_\mu p_\mu - m)u = 0.$$

There is no natural choice between positive and negative energy-free particle solutions to the Dirac equation; all we know is that the energy should satisfy the relation $E^2 - p^2 = m^2$. Taken with the two possible spin alignments there are in all four distinguishable electron states for a given three-momentum: spin up, $E>0$; spin down, $E>0$; spin up, $E<0$; spin down, $E<0$. u is thus a four-component entity and the γ's which act on it are 4×4 matrices. It is helpful to consider first the familiar non-relativistic case in which the spin up and spin down states are represented by two-component entities called spinors:

$$\phi_+ = \begin{bmatrix} 1 \\ 0 \end{bmatrix} \quad \phi_- = \begin{bmatrix} 0 \\ 1 \end{bmatrix}$$

and the angular momentum operator is:

$$\mathbf{s} = (\hbar/2)\boldsymbol{\sigma}.$$

$\boldsymbol{\sigma}$ is the Pauli spin matrix with components

$$\sigma_1 = \begin{bmatrix} 0 & 1 \\ 1 & 0 \end{bmatrix} \quad \sigma_2 = \begin{bmatrix} 0 & -i \\ i & 0 \end{bmatrix} \quad \text{and} \quad \sigma_3 = \begin{bmatrix} 1 & 0 \\ 0 & -1 \end{bmatrix}.$$

Then $s_3\phi_\pm = \pm(\hbar/2)\phi_\pm$ and $s^2\phi_\pm = (3\hbar^2/4)\phi_\pm$.

In the relativistic case the forms of the γ matrices are needed:

$$\gamma_0 = \begin{bmatrix} I_2 & 0 \\ 0 & -I_2 \end{bmatrix} \quad \gamma_i = \begin{bmatrix} 0 & \sigma_i \\ -\sigma_i & 0 \end{bmatrix},$$

where each 0 represents a 2×2 null matrix $\begin{bmatrix} 0 & 0 \\ 0 & 0 \end{bmatrix}$ and each I_2 a 2×2 diagonal unit matrix $\begin{bmatrix} 1 & 0 \\ 0 & 1 \end{bmatrix}$. The positive-energy four-component spinors have the form

$$u(p, s) = \sqrt{E+m} \begin{bmatrix} \phi_s \\ \dfrac{\boldsymbol{\sigma}\cdot\mathbf{p}}{(E+m)}\phi_s \end{bmatrix},$$

where $\phi_s = \phi_\pm$ for $s = \pm\frac{1}{2}$ respectively. We also need the notation

$$u = \begin{bmatrix} u_1 \\ u_2 \\ u_3 \\ u_4 \end{bmatrix}.$$

The analogue of the conjugate of a vector (row vector) is the adjoint of the spinor

$$\bar{u} = u^+\gamma_0 = [\bar{u}_1\ \bar{u}_2\ \bar{u}_3\ \bar{u}_4],$$

where u^+ is the complex conjugate transpose of u. With these definitions the spin states are orthogonal

$$\bar{u}_\alpha(p, +\tfrac{1}{2})u_\alpha(p, -\tfrac{1}{2}) = 0$$

and normalized such that

$$\bar{u}_\alpha(p, s)u_\alpha(p, s) = 2m,$$

where the implied sum is made over the spinor index α running from 1 to 4. It is usual to drop the spinor indices for simplicity, so these relations are written as

$$\bar{u}(p, \pm\tfrac{1}{2})u(p, \pm\tfrac{1}{2}) = 0, \quad \bar{u}(p, s)u(p, s) = 2m$$

or more briefly

$$\bar{u}(p, s)u(p, r) = 2m\delta_{rs}.$$

Positrons are described by the spinors

$$v(p, s) = \sqrt{E+m} \begin{bmatrix} \dfrac{\boldsymbol{\sigma} \cdot \mathbf{p}}{(E+m)} \chi_s \\ \chi_s \end{bmatrix},$$

where $\chi_+ = \begin{bmatrix} 0 \\ 1 \end{bmatrix} = \phi_-, \quad \chi_- = \begin{bmatrix} 1 \\ 0 \end{bmatrix} = \phi_+.$

The adjoint is $\bar{v} = v^+\gamma^0$ and the normalization has $\bar{v}v = -2m$.

The wavefunction of a positron of positive energy E, momentum $\mathbf{p}$ and spin component s is

$$\psi(p, s) = v(p, s) \exp(-ip_\mu x_\mu).$$

In the particle rest frame the four solutions to the Dirac equation reduce to

$$\text{electron spin } \pm\tfrac{1}{2} \quad u(0, \pm) = \sqrt{2m} \begin{bmatrix} \phi_\pm \\ 0 \\ 0 \end{bmatrix}$$

$$\text{positron spin } \pm\tfrac{1}{2} \quad v(0, \pm) = \sqrt{2m} \begin{bmatrix} 0 \\ 0 \\ \chi_\pm \end{bmatrix}$$

showing the expected collapse into non-relativistic two-component spinors. The positive- and negative-energy spinors are orthogonal, so it is possible to write down an operator which is useful in projecting out positive states

$$(\Lambda_+)_{\alpha\beta} = \sum_r u_\alpha(p, r)\bar{u}_\beta(p, r),$$

which is usually written with the spinor subscripts suppressed,

$$\Lambda_+ = \sum_r u(p, r)\bar{u}(p, r).$$

In addition, again leaving out spinor subscripts,

$$(\gamma_\mu p_\mu + m)u(p, s) = 2mu(p, s)$$

and

$$(\gamma_\mu p_\mu + m)v(p, s) = 0$$

so that

$$\Lambda_+ = (\gamma_\mu p_\mu + m) = \sum_r u(p, r)\bar{u}(p, r),$$

a relation which will be much used in simplifying matrix elements.

Neutrinos are also spin-$\frac{1}{2}$ fermions but unlike the electrons possess no mass and no charge. Starting with a four-component spinor $\begin{bmatrix} w_1 \\ w_2 \end{bmatrix}$ we first project out a piece using the operator

$$P_L = \tfrac{1}{2}\begin{bmatrix} I_2 & -I_2 \\ -I_2 & I_2 \end{bmatrix}$$

to get

$$\begin{bmatrix} \tfrac{1}{2}(w_1 - w_2) \\ \tfrac{1}{2}(w_2 - w_1) \end{bmatrix}.$$

Substituting this into the Dirac equation with m set to zero gives

$$\gamma_0 \frac{\partial}{\partial t}\begin{bmatrix} w_1 - w_2 \\ w_2 - w_1 \end{bmatrix} = -\boldsymbol{\gamma}\cdot\nabla\begin{bmatrix} w_1 - w_2 \\ w_2 - w_1 \end{bmatrix}$$

i.e.

$$\frac{\partial}{\partial t}\begin{bmatrix} w_1 - w_2 \\ w_1 - w_2 \end{bmatrix} = \begin{bmatrix} \boldsymbol{\sigma}\cdot\nabla & (w_1 - w_2) \\ \boldsymbol{\sigma}\cdot\nabla & (w_1 - w_2) \end{bmatrix}.$$

Therefore the massless neutrino only requires a *two*-component spinor $(w_1 - w_2)$ for its description. We call this $w(\nu)$, and if the neutrino four-momentum is $(E, \mathbf{p})$,

$$Ew(\nu) = -\boldsymbol{\sigma}\cdot\mathbf{p}w(\nu).$$

From the above equation the neutrino is seen to have a spin component $\boldsymbol{\sigma}\cdot\mathbf{p}/E$ of -1. This is the spin component projected along the direction of motion of the neutrino and is called the helicity (λ). The neutrino is therefore *left-handed*. An anti-neutrino will also require only a two-component spinor which is orthogonal to $w(\nu)$, namely $w(\nu) = w_1 + w_2$, which is *right-handed*:

$$Ew(\bar{\nu}) = \boldsymbol{\sigma}\cdot\mathbf{p}w(\bar{\nu}).$$

The operator P_L projects out the left-handed component:

$$P_L w(\nu) = w(\nu), \quad P_L w(\bar{\nu}) = 0.$$

P_L is conventionally written

$$P_L = \tfrac{1}{2}(1 - \gamma_5),$$

where in the representation of γ matrices used here

$$\gamma_5 = \begin{bmatrix} 0 & I_2 \\ I_2 & 0 \end{bmatrix}.$$

The corresponding right-handed projection operator is

$$P_R = \tfrac{1}{2}(1 + \gamma_5).$$

A general definition of γ_5, true for any representation of the γ matrices, is

$$\gamma_5 = i\gamma_0\gamma_1\gamma_2\gamma_3.$$

As presented here the choice of the neutrino as left-handed and the antineutrino as right-handed is quite arbitrary. The neutrino could just as well have been chosen to be right-handed. It is only experiment that can decide between the possibilities, and in section 7.2 the evidence that the neutrino is left-handed is presented.

Appendix E
The electromagnetic cross-section for $\frac{1}{2}^+ \frac{1}{2}^+ \rightarrow \frac{1}{2}^+ \frac{1}{2}^+$

Single-photon exchange dominates elastic electromagnetic processes and as an example we calculate the amplitude for elastic point-like fermion scattering. This will prove useful as a prototype for other calculations where single-boson exchange is important, such as deep inelastic neutrino scattering. In the reaction $1+2\rightarrow 3+4$, where 1, 2, 3 and 4 are spin-$\frac{1}{2}$ point-like fermions the Feynman amplitude obtained using the rules given in Chapter 4 is:

$$T=[\bar{u}(k_3, s_3)\gamma_\mu u(k_1, s_1)]\frac{e^2}{q^2}[\bar{u}(k_4, s_4)\gamma_\mu u(k_2, s_2)],$$

where k is the particle momentum four-vector and s is the spin component. The exchanged momentum, q, is (k_1-k_3). The cross-section is given by the standard form described in Appendix A:

$$d\sigma=A\ dQ_2/F,$$

where A contains an average over initial-state spin components and a sum over final-state spin components:

$$A=\tfrac{1}{4}\sum_{\text{spin}}|T|^2.$$

T is of the form $X\dfrac{e^2}{q^4}Y$, so that T^* is $Y^+\dfrac{e^2}{q^4}X^+$. Thus

$$T^*=[u^+(k_2, s_2)\gamma_\nu^+\gamma_0 u(k_4,s_4)]$$
$$\frac{e^2}{q^4}[u^+(k_1, s_1)\gamma_\nu^+\gamma_0 u(k_3, s_3)],$$

where we have used the equalities

$$(\bar{u})^+ = (u^+\gamma_0)^+ = \gamma_0^+(u^+)^+ = \gamma_0 u.$$

Thus

$$A = (e^4/4q^4)\sum_s$$

$$[u^+(k_2, s_2)\gamma_\nu^+\gamma_0 u(k_4, s_4)u^+(k_1, s_1)\gamma_\nu^+\gamma_0 u(k_3, s_3)]$$
$$[\bar{u}(k_3, s_3)\gamma_\mu u(k_1, s_1)\bar{u}(k_4, s_4)\gamma_\mu u(k_2, s_2)].$$

Thus

$$A = (e^4/4q^4)L_{\nu\mu}(2, 4)L_{\nu\mu}(1, 3),$$

where

$$L_{\nu\mu}(2, 4) = \sum_{s_2 s_4} \bar{u}(k_2, s_2)\gamma_\nu u(k_4, s_4)\bar{u}(k_4, s_4)\gamma_\mu u(k_2, s_2),$$

with a similar expression for $L_{\mu\nu}(1, 3)$. We have used here the equality $\gamma_0\gamma_\nu^+\gamma_0 = \gamma_\nu$. From results given in the previous appendix the spin sums are

$$\sum_{s_4} u_\alpha(k_4, s_4)\bar{u}_\beta(k_4, s_4) = (\gamma \cdot k_4 + m_2)_{\alpha\beta}$$

$$\sum_{s_2} u_\gamma(k_2, s_2)\bar{u}_\delta(k_2, s_2) = (\gamma \cdot k_2 + m_2)_{\gamma\delta},$$

where the four-component spinor subscripts α, β, δ and γ have been written explicitly. $\gamma \cdot k_4$ means the scalar product of the space-time four-vectors γ and k_4. In a bit more detail

$$(\gamma \cdot k)_{\alpha\beta} = (\gamma_0)_{\alpha\beta}E - (\boldsymbol{\gamma})_{\alpha\beta} \cdot \mathbf{k}.$$

m_2 is understood to mean $m_2 I_4$, where I_4 is the 4×4 unit matrix for spinors. Thus

$$L_{\nu\mu}(2, 4) = (\gamma_\nu)_{\delta\alpha}(\gamma \cdot k_4 + m_2)_{\alpha\beta}(\gamma_\mu)_{\beta\gamma}(\gamma \cdot k_2 + m_2)_{\gamma\delta}.$$

Noting that the initial and final spinor indices are identical:

$$L_{\nu\mu}(2, 4) = \text{Trace}[\gamma_\nu(\gamma \cdot k_4 + m_2)\gamma_\mu(\gamma \cdot k_2 + m_2)].$$

At this point the following trace theorems for the γ matrices are useful:

$$\text{Trace}\,(\gamma_\nu\gamma_\mu)=(\gamma_\nu)_{\alpha\beta}(\gamma_\mu)_{\beta\alpha}=4g_{\nu\mu},$$
$$\text{Trace}\,(\gamma_\nu\gamma_\rho\gamma_\mu\gamma_\sigma)=4[g_{\nu\rho}g_{\mu\sigma}-g_{\nu\mu}g_{\rho\sigma}+g_{\nu\sigma}g_{\rho\mu}],$$

while traces of products of odd numbers of γ matrices are zero. Then

$$L_{\nu\mu}(2,4)=4[k_{4\nu}k_{2\mu}+k_{4\mu}k_{2\nu}-g_{\nu\mu}(k_4\cdot k_2-m_2^2)].$$

When taking the product of $L(2, 4)$ and $L(1, 3)$ we make use of the result $g_{\mu\nu}g_{\mu\nu}=4$. Finally the amplitude term is

$$A=(8e^4/q^4)[(k_4\cdot k_3)(k_2\cdot k_1)+(k_4\cdot k_1)(k_2\cdot k_3)$$
$$-m_1^2(k_4\cdot k_2)-m_2^2(k_1\cdot k_3)+2m_1^2m_2^2],$$

a *master equation* which will be used several times. As a first example let us calculate the cross-section for $e^-+e^+\rightarrow\mu^-+\mu^+$ in the CM frame. Figure 4.3 compares the two processes; (a) $e^-+e^+\rightarrow\mu^-+\mu^+$ and (b) $e^-+\mu^-\rightarrow e^-+\mu^-$. Both involve a single-photon intermediate state; in one case as an exchanged particle and in the other, (a) it is a virtual particle in what is called the direct channel. Therefore the two processes have the same *analytic form* for their amplitudes, though not the same kinematic factors. We specialize to the case where the energy is sufficiently large that all masses may be neglected. Then for reaction (a)

$$A=(8e^4/q^4)[(k_4\cdot k_3)(k_2\cdot k_1)+(k_4\cdot k_1)(k_2\cdot k_3)]$$
$$=(8e^4/q^4)[(\bar{\mu}\cdot\bar{e})(\mu\cdot e)+(\bar{\mu}\cdot e)(\mu\cdot\bar{e})],$$

where we have written the particle names to denote their four-momenta. We evaluate the cross-section in the CM frame. If the masses are negligible all the particles have identical energies and momenta in this frame. Thus

$$\bar{\mu}\cdot\bar{e}=\mu\cdot e=\mu_0 e_0-\boldsymbol{\mu}\cdot\mathbf{e}$$
$$=E^2(1-\cos\theta^*),$$

where θ^* is the angle the μ^- direction makes with respect to the incoming e^-. Similarly:

$$\mu\cdot\bar{e}=\bar{\mu}\cdot e=E^2(1+\cos\theta^*)$$

because $(\pi-\theta^*)$ is the angle the μ^+ direction makes with the incoming e^-. Therefore

$$A=(8e^4E^4/q^4)[(1+\cos\theta^*)^2+(1-\cos\theta^*)^2].$$

Notice that for the reaction (a) the photon four-momentum squared, q^2, is also the total energy squared, i.e. $(2E)^2$. Thus

$$A=e^4(1+\cos^2\theta^*).$$

From Appendix A the kinematic factors in the CM frame are:

$$\mathrm{d}Q_2=\mathrm{d}\Omega^*/32\pi^2 \quad \text{and} \quad F=2s$$

for the case that masses are negligible. Therefore

$$\begin{aligned}\mathrm{d}\sigma/\mathrm{d}\Omega&=(e^4/64\pi^2 s)(1+\cos^2\theta^*)\\&=(\alpha^2/4s)(1+\cos^2\theta^*).\end{aligned}$$

Integrating over all directions gives the total cross-section:

$$\sigma=4\pi\alpha^2/3s.$$

These results can be taken over for the reaction $e^++e^-\rightarrow q+\bar{q}$ if the muon charge is replaced by the quark charge.

The weak cross-section for $\frac{1}{2}^+\ \frac{1}{2}^+\rightarrow\frac{1}{2}^+\ \frac{1}{2}^+$

The amplitude for the underlying weak processes has a current times current structure and is discussed in Chapter 7. In the following the cross-sections for lepton-lepton and lepton-quark scattering are calculated. The first example considered is:

$$\begin{aligned}&\nu_\mu+e^-\rightarrow\mu^-+\nu_e\\&(1+2\rightarrow3+4).\end{aligned}$$

The amplitude is

$$\begin{aligned}T=(G/\sqrt{2})&[\bar{u}(k_3,s_3)\gamma_\mu(1-\gamma_5)u(k_1,s_1)]\\&[\bar{u}(k_4,s_4)\gamma_\mu(1-\gamma_5)u(k_2,s_2)]\end{aligned}$$

and the cross-section is given by

$$\mathrm{d}\sigma=A\,\mathrm{d}Q_2/F,$$

where A includes an average over initial spin state contributions and a sum over final-state spins:

$$A = \tfrac{1}{2}\sum_s |T|^2.$$

Neutrinos only have one spin component, so the spin average leads to a factor $\frac{1}{2}$. The term A is the product of a muon and an electron tensor:

$$A = (G^2/4)M_{\nu\mu}(1, 3)E_{\nu\mu}(2, 4).$$

We can use the expression obtained earlier in this appendix for $L_{\nu\mu}(1, 3)$ as a guide to the form and the reduction of these two tensors. Then

$$M_{\nu\mu}(1, 3) = \sum_{s_1, s_3} [\bar{u}(k_1, s_1)\gamma_\nu(1-\gamma_5)u(k_3, s_3)]$$
$$\times [\bar{u}(k_3, s_3)\gamma_\mu(1-\gamma_5)u(k_1, s_1)].$$

We specialize to the case that the energies are large enough for all masses to be neglected. From Appendix D we use the spin sum

$$\sum_s u(k, s)\bar{u}(k, s) = k \cdot \gamma.$$

Therefore

$$M_{\nu\mu}(1, 3) = \text{Trace}\,[(k_1 \cdot \gamma)\gamma_\nu(1-\gamma_5)(k_3 \cdot \gamma)\gamma_\mu(1-\gamma_5)].$$

We note that γ_5 commutes with an even number of γ matrices. Thus

$$M_{\nu\mu}(1, 3) = \text{Trace}\,[(1-\gamma_5)^2(k_1 \cdot \gamma)\gamma_\nu(k_3 \cdot \gamma)\gamma_\mu]$$
$$= 2\,\text{Trace}\,[(1-\gamma_5)(k_1 \cdot \gamma)\gamma_\nu(k_3 \cdot \gamma)\gamma_\mu],$$

where we have used the equality $(1-\gamma_5)^2 = 2(1-\gamma_5)$. With the help of the trace theorems given above this further reduces to

$$M_{\nu\mu}(1, 3) = 8[k_{1\nu}k_{3\mu} + k_{1\mu}k_{3\nu} - g_{\nu\mu}k_1 \cdot k_3$$
$$+ i\varepsilon_{\alpha\nu\beta\mu}k_{1\alpha}k_{3\beta}].$$

In particular the evaluation of this last term has made use of

$$\text{Trace}\,[\gamma_5\gamma_\alpha\gamma_\nu\gamma_\beta\gamma_\mu] = -4i\varepsilon_{\alpha\nu\beta\mu}$$

where ε is asymmetric under interchange of adjacent subscripts and $\varepsilon_{0123} = +1$. The last term in $M_{\nu\mu}$ flips sign under the parity transformation and was absent in the electromagnetic case. It leads to parity-violating effects. The product $M_{\nu\mu}E_{\nu\mu}$ is an intensity, so that all imaginary terms in the product cancel. This leaves the product of the two real terms, which was evaluated earlier in this Appendix, and the product of the imaginary terms.

$$\text{Real} \times \text{Real} = 128[(k_1 \cdot k_2)(k_3 \cdot k_4) + (k_1 \cdot k_4)(k_2 \cdot k_3)].$$

$$\begin{aligned}\text{Imag} \times \text{Imag} &= 64[\varepsilon_{\alpha\nu\beta\mu}k_{1\alpha}k_{3\beta}\varepsilon_{\gamma\nu\delta\mu}k_{2\gamma}k_{4\delta}] \\ &= 128[(k_1 \cdot k_2)(k_3 \cdot k_4) - (k_1 \cdot k_4)(k_2 \cdot k_3)].\end{aligned}$$

The final result is quite simple

$$M_{\nu\mu}E_{\nu\mu} = 256(k_1 \cdot k_2)(k_3 \cdot k_4).$$

The cross-section for neutrino-lepton charged-current scattering is then

$$\mathrm{d}\sigma(\nu) = (G^2/4)256(k_1 \cdot k_2)(k_3 \cdot k_4)\,\mathrm{d}Q_2/F.$$

The expressions appropriate for the CM frame are to be found in Appendix A:

$$\mathrm{d}Q_2 = \mathrm{d}\Omega^*/32\pi^2, \quad F = 2s.$$

Therefore

$$\mathrm{d}\sigma/\mathrm{d}\Omega^* = (G^2/\pi^2 s)(k_1 \cdot k_2)(k_3 \cdot k_4),$$

where s is the CM energy squared. In the approximation that all masses are negligible we have, in the CM frame,

$$k_1 \cdot k_2 = (k_1 + k_2)^2/2 = s/2 \quad \text{and} \quad k_3 \cdot k_4 = s/2.$$

Therefore

$$\mathrm{d}\sigma/\mathrm{d}\Omega^* = G^2 s/4\pi^2,$$

which is isotropic. The total cross-section for $\nu_\mu + e^- \rightarrow \mu^- + \nu_e$ is thus $G^2 s/\pi$ and would rise *indefinitely* as the energy increases.

The development given above can also be used to obtain cross-sections for the neutrino-quark charged-current processes considered in Chapter 9:

$$\nu_\mu + \mathrm{d} \rightarrow \mu^- + \mathrm{u} \qquad [E.1]$$

and

$$\nu_\mu + \bar{\mathrm{u}} \rightarrow \mu^- + \bar{\mathrm{d}}. \qquad [E.2]$$

In fact, we can take over the results given immediately above for reaction [E.1], which is a fermion-fermion scattering. Explicitly

$$d\sigma = 64G^2(k_1 \cdot k_2)(k_3 \cdot k_4)\, dQ_2/F,$$

so that the angular distribution in the neutrino-quark CM frame is:

$$d\sigma/d\Omega^* = G^2 s/4\pi^2.$$

Reaction (2) can be obtained from section (1) by interchange of particles 2 and 4. Thus the cross-section

$$d\sigma = 64G^2(k_1 \cdot k_4)(k_2 \cdot k_3)\, dQ_2/F.$$

The cross-section will be evaluated in the CM frame. If θ^* is the angle through which the lepton scatters, then

$$\mathbf{k}_1 \cdot \mathbf{k}_4 = \mathbf{k}_2 \cdot \mathbf{k}_3 = -(s/4)\cos\theta^*$$

and

$$k_1 \cdot k_4 = k_2 \cdot k_3 = (s/4)(1 + \cos\theta^*).$$

Thus

$$d\sigma = 4G^2 s^2 (1 + \cos\theta^*)^2\, dQ_2/F.$$

Then the angular distribution for reaction (2) is

$$d\sigma/d\Omega^* = (G^2 s/16\pi^2)(1 + \cos\theta^*)^2.$$

Appendix F
Exercises and worked solutions

Exercises

Q2.1 Show that the four-momentum transfer squared $(-q^2)$ for elastic scattering through a small angle θ is approximately equal to the transverse momentum squared of the scattered particle $p_T^2 = (p \sin \theta)^2$. If the width of the target is d show that the width of the $-q^2$ distribution is of order d^{-2}.

Q2.2 In the following list only one example is a valid weak decay; the remainder all violate some conservation law. Which is the valid example and why are the others invalid?

(a) $\Lambda(1115) \rightarrow p(939) + \pi^0(135)$
(b) $\Lambda^0(1115) \rightarrow p(939) + \pi^-(140) + \pi^0(135)$
(c) $\Xi^0(1315) \rightarrow p(939) + \pi^-(140)$
(d) $\Lambda^0(1116) \rightarrow \pi^+(140) + \pi^-(140)$
(e) $\Xi^0(1315) \rightarrow \Lambda^0(1116) + \pi^0(135)$

Q3.1 Calculate the momenta above which π-mesons, K-mesons and protons begin to emit Cerenkov radiation when travelling through air (the refractive index of air is 1·00027). Over what range of momenta would an air-filled threshold Cerenkov counter be useful in discriminating π-mesons from K-mesons?

Q3.2 A charged particle of momentum p GeV/c travels a distance l metres in a bubble chamber in a plane perpendicular to the applied magnetic field of B teslas. Calculate the sagitta of the track. Suppose the bubble chamber is photographed at a demagnification of 20 with a camera whose optic axis is parallel to the field. The coordinates of the end points and a point near the middle of the track are measured on the film. Calculate the uncertainty in the momentum determination due to measurement error if the field is 1 tesla, the track is 1 m long, and the measurement accuracy is 5 μm in the film plane.

Q4.1 Show, starting from the forms given in Appendix C, that

(a) $\gamma_5 = i\gamma_0\gamma_1\gamma_2\gamma_3 = \begin{bmatrix} \cdot & I_2 \\ I_2 & \cdot \end{bmatrix}$

(b) $P_L = \frac{1}{2}\begin{bmatrix} I_2 & -I_2 \\ -I_2 & I_2 \end{bmatrix}, \quad P_R = \frac{1}{2}\begin{bmatrix} I_2 & I_2 \\ I_2 & I_2 \end{bmatrix}$

(c) $P_L P_R = 0, \quad P_L P_L = P_L$

Q4.2 The trace (sum of diagonal elements) of $\gamma_\mu\gamma_\nu$ is $(\gamma_\mu)_{\alpha\beta}(\gamma_\nu)_{\beta\alpha}$ with summation over the spinor components α and β. Prove that Trace $(\gamma_\mu\gamma_\nu) = 4g_{\mu\nu}$.

Q4.3 Calculate the total cross-section for $e^+ + e^- \rightarrow \mu^+ + \mu^-$ at CM energies of 2 GeV and 40 GeV. What are the corresponding cross-sections for $e^+ + e^- \rightarrow \tau^+ + \tau^-$?

Q5.1 A U(1) Lie group can be formed with the momentum component p_x as the generator. Show that the group operations in space are translations along the x-direction. Hence show that if all physical processes are invariant under translations, p_x is a conserved quantity.

Q5.2 The π^0-meson decays predominantly to two photons. Starting from the assumption that the parent has spin zero show that the polarization state of the photons is $LL' \pm RR'$ if the parent has even/odd parity. R (L) stands for a right (left) circularly polarized photon.

Q6.1 Calculate the effect of $F_1 \pm iF_2$, $F_4 \pm iF_5$ and $F_6 \pm iF_7$ acting on each of the three quark states. The explicit forms of F_i differ between the quark and antiquark representation: $F_1(\bar{3}) = -[F_i(3)]^*$. Check that the effects on the antiquark states are consistent. Finally show that all the generators annihilate $(r\bar{r}+y\bar{y}+b\bar{b})/\sqrt{3}$, and so prove that it is a colour singlet state.

Q6.2 Calculate α_s for $-q^2$ equal to $(10^3\ \text{GeV})^2$ and $(10^{15}\ \text{GeV})^2$.

Q7.1 (a) Show that the neutral K-meson decays to $\pi^- + e^+ + \nu_e$ and $\pi^+ + e^- + \bar{\nu}_e$ come from parents of opposite strangeness. (b) Explain why the decay $D^+ \to K^- + \pi^+ + \pi^+$ has a branching fraction $4{\cdot}6 \pm 1{\cdot}1$ per cent and yet the decay $D^+ \to K^+ + \pi^+ + \pi^-$ is barely detected.

Q7.2 (a) The angular distribution for $\nu_\mu + e^- \to \mu^- + \nu_e$ is isotropic in the CM frame. There is an upper limit to s-wave cross-section of $2\pi/s$ which is reached when the incoming s-wave component of the ν_μ beam is fully scattered. Any higher cross-section would correspond to having more μ^-'s outgoing than there were ν_μ's incident. Calculate the CM energy at which this 'unitarity bound' is reached by the Fermi cross-section. (b) Show that in the laboratory the maximum angle the μ^- can make with the incident ν_μ direction is $\sqrt{2m_e/E_\nu}$, where E_ν is the neutrino energy in the laboratory frame and is very much greater than any of the masses involved.

Q8.1 Explain why both elastic scattering of ν_e or $\bar{\nu}_e$ from electrons is possible through the charged-current interaction but not elastic scattering of ν_μ or $\bar{\nu}_\mu$.

Q8.2 (a) Show that the decay rates of the Z^0 are in the proportions

$$\Gamma(\nu_e\bar{\nu}_e):\Gamma(e^+e^-):\Gamma(u\bar{u}):\Gamma(d\bar{d})$$
$$= 1:8\sin^4\theta_W - 4\sin^2\theta_W + 1:3[32/9\sin^4\theta_W - 8/3\sin^2\theta_W + 1]$$
$$:3[8/9\sin^4\theta_W - 4/3\sin^2\theta_W + 1]$$

(b) What are the numerical branching ratios for $Z^0 \to \nu_e\bar{\nu}_e$ and

$Z^0 \rightarrow e^+e^-$? (c) If the partial width (decay rate) for $Z^0 \rightarrow \nu_e\bar{\nu}_e$ is $GM_z^3/12\pi\sqrt{2}$ calculate the total width in GeV/c^2.

Q9.1 An electron of 200 GeV energy is scattered inelastically from a proton and emerges at 1° from its initial direction with an energy of 160 GeV. Calculate the four-momentum transfer squared. What is the x value of the target parton? What is the probability that this parton is a quark rather than an antiquark?

Q9.2 Show that the cross-section for production of relatively massive Drell-Yan pairs in $\bar{p}p$ collisions reduces to

$$[\mathrm{d}^2\sigma/\mathrm{d}x_F\,\mathrm{d}M]_{x_f=0} = (4\pi K\alpha^2/9W^3\tau)\,\Sigma Q_f^2 v_f^2(\sqrt{\tau}),$$

where W is the CM energy and v_f is the proton valence quark structure function.

Q10.1 The $\psi'(3685)$ can decay to $\psi(3097)+\pi^+ + \pi^-$, but the partial width is only 70 keV. Can you explain why this decay is inhibited?

Q10.2 (a) Most members of the baryon octet have quark content $q_1q_1q_2$. Show that the spin state of such a baryon with spin up is $[\sqrt{2/3}|1, +1\rangle_1|\tfrac{1}{2}, -\tfrac{1}{2}\rangle_2 - \sqrt{1/3}|1, 0\rangle_1|\tfrac{1}{2}, +\tfrac{1}{2}\rangle_2]$. Hence show that their magnetic moments are $[(2/3)(2\mu(1)-\mu(2))+(1/3)\mu(2)]$, where $\mu(1)$ and $\mu(2)$ are the magnetic moments of the quarks q_1 and q_2 respectively. (b) Show that $\mu(u)=0{\cdot}67\mu$, $\mu(d)=-0{\cdot}33\mu$ and $\mu(s)=-0{\cdot}23\mu$, where $\mu=(ge/2m_u)$ and g is the quark gyromagnetic ratio. Hence from (a) show that μ equals the proton magnetic moment (2·79 nuclear magnetons). (c) Make predictions for the magnetic moments (in nuclear magnetons) of the neutron, Σ^+, Σ^-, Ξ^0 and Ξ^- hyperons. (d) show that the Λ-hyperon magnetic moment is μ_s.

Q10.3 Show that at the CM energy corresponding to the $\Delta(1232)$ resonance the cross-sections are in the ratios

$\pi^+ + p \rightarrow \pi^+ + p$ 1

$\pi^- + p \rightarrow \pi^0 + n$ 2/9

$\pi^- + p \rightarrow \pi^- + p$ 1/9

$\pi^- + n \rightarrow \pi^- + n$ 1

Q11.1 Show that the distribution $N(p_L)\,dp_L \propto dp_L/E$ translates into a flat distribution in rapidity.

Q11.2 The ρ-meson has a Regge trajectory given by $\alpha_\rho(t) = 0{\cdot}55 + 0{\cdot}9t$. Calculate the reduction in the forward differential cross-section for $\pi^- p$ charge exchange on going from a beam momentum of 20 GeV/c to 200 GeV/c.

Q11.3 Show that according to the quark model the reactions $K^- + p \rightarrow \Lambda^0 + \rho^0$ and $K^- + p \rightarrow \Lambda^0 + \omega^0$ have equal cross-sections.

Solutions

A2.1 Taking results from section 2.3 the four-momentum transfer squared

$$-q^2 = 4p^2 \sin^2(\theta/2)$$
$$\simeq p^2 \sin^2\theta \quad \text{for } \theta \text{ small}$$
$$= p_T^2.$$

If the lateral dimension is d then the angular size of the diffraction pattern is

$$\theta \simeq 1{\cdot}2\lambda/d$$

for wavelength λ.

de Broglie's relation is $p = \lambda^{-1}$ $(\hbar = c = 1)$. Thus

$$p\theta \simeq 1{\cdot}2d^{-1}$$

i.e.

$$p_T \simeq d^{-1}.$$

Thus

$$-q^2 \simeq d^{-2}.$$

A2.2 (a) Forbidden because charge is not conserved. (b) Forbidden energetically because the sum of masses of the products exceeds the mass of the parent. (c) Forbidden because there is a change of strangeness of two units. (d) Forbidden because the number of baryons changes. (e) This is the dominant mode ($\sim$100 per-cent) for $\Xi^0(1315)$ decay.

A3.1 Charged particles begin to emit Cerenkov radiation when their velocity βc exceeds the local velocity of light, i.e. c/n, where n is the refractive index. In the case of air the threshold is

$$\beta = 0{\cdot}99973,$$
$$\gamma = 43{\cdot}042.$$

For a particle of mass the threshold momentum is $m\beta\gamma$, i.e. $43{\cdot}0303m$. The masses of π^+, K^+ and p are 139·57, 493·67 and 938·28 MeV/c^2, so that the corresponding threshold momenta are 6·01 GeV/c, 21·24 GeV/c and 40·37 GeV/c.

Below 6·01 GeV/c neither π-mesons nor K-mesons would produce Cerenkov radiation and above 21·24 GeV/c both species produce Cerenkov radiation. But between these two momenta (6·01 and 21·24 GeV/c) only the π-mesons would radiate and so could be distinguished from K-mesons.

A3.2 The radius of curvature ρ (metres) of the track is given by

$$p = 0{\cdot}3B\rho$$

i.e.

$$\rho = p/0{\cdot}3B.$$

Then the sagitta

$$s = l^2/8\rho = 0{\cdot}3l^2B/8p.$$

Thus,

$$p = 0{\cdot}3l^2B/8s$$

and

$$\begin{aligned}\Delta p/p &= \Delta s/s \\ &= 8p\Delta s/0{\cdot}3l^2B.\end{aligned}$$

If $l = 1$ m and $B = 1$ T and Δs is $20 \times 5\ \mu\text{m} = 100\ \mu\text{m}$,

$$\Delta p/p = 0{\cdot}0022p \text{ (in GeV}/c\text{)}.$$

A4.1 (a)

$$\begin{aligned}\gamma_0\gamma_1 &= \begin{bmatrix} I_2 & \cdot \\ \cdot & -I_2 \end{bmatrix}\begin{bmatrix} 0 & \sigma_1 \\ -\sigma_1 & 0 \end{bmatrix} \\ &= \begin{bmatrix} \cdot & \sigma_1 \\ \sigma_1 & \cdot \end{bmatrix}\end{aligned}$$

$$\begin{aligned}\gamma_2\gamma_3 &= \begin{bmatrix} \cdot & \sigma_2 \\ -\sigma_2 & \cdot \end{bmatrix}\begin{bmatrix} \cdot & \sigma_3 \\ -\sigma_3 & \cdot \end{bmatrix} \\ &= \begin{bmatrix} -\sigma_2\sigma_3 & \cdot \\ \cdot & -\sigma_2\sigma_3 \end{bmatrix}.\end{aligned}$$

Therefore

$$i\gamma_0\gamma_1\gamma_2\gamma_3 = -i\begin{bmatrix} \cdot & \sigma_1\sigma_2\sigma_3 \\ \sigma_1\sigma_2\sigma_3 & \cdot \end{bmatrix}$$

$$\begin{aligned}\sigma_2\sigma_3 &= \begin{bmatrix} \cdot & -i \\ i & \cdot \end{bmatrix}\begin{bmatrix} 1 & \cdot \\ \cdot & -1 \end{bmatrix} \\ &= \begin{bmatrix} \cdot & i \\ i & \cdot \end{bmatrix}.\end{aligned}$$

Therefore

$$\sigma_1\sigma_2\sigma_3 = \begin{bmatrix} \cdot & 1 \\ 1 & \cdot \end{bmatrix}\begin{bmatrix} \cdot & i \\ i & \cdot \end{bmatrix} = \begin{bmatrix} i & \cdot \\ \cdot & i \end{bmatrix} = iI_2.$$

Therefore

$$i\gamma_0\gamma_1\gamma_2\gamma_3 = \begin{bmatrix} \cdot & I_2 \\ I_2 & \cdot \end{bmatrix}.$$

(b)

$$p_L = \tfrac{1}{2}(1-\gamma_5) = \tfrac{1}{2}\begin{bmatrix} I_2 & -I_2 \\ -I_2 & I_2 \end{bmatrix}$$

$$= \begin{bmatrix} 1 & \cdot & -1 & \cdot \\ \cdot & 1 & \cdot & -1 \\ -1 & \cdot & 1 & \cdot \\ \cdot & -1 & \cdot & 1 \end{bmatrix}$$

and

$$p_R = \tfrac{1}{2}\begin{bmatrix} I_2 & I_2 \\ I_2 & I_2 \end{bmatrix}.$$

(c)

Then

$$p_L p_R = \tfrac{1}{4}\begin{bmatrix} I_2 - I_2 & I_2 - I_2 \\ I_2 - I_2 & I_2 - I_2 \end{bmatrix} = 0$$

and

$$p_L p_L = \tfrac{1}{4}\begin{bmatrix} I_2 + I_2 & -I_2 - I_2 \\ -I_2 - I_2 & I_2 + I_2 \end{bmatrix} = p_L.$$

A4.2 Note that the components $(\gamma_\mu)_{\alpha\beta}$ and $(\gamma_\nu)_{\beta\alpha}$ of the 4×4 matrices are pure numbers and hence their order may be reversed without affecting the product. Therefore

$$\text{Trace } (\gamma_\mu \gamma_\nu) = (\gamma_\mu)_{\alpha\beta}(\gamma_\nu)_{\beta\alpha}$$
$$= (\gamma_\nu)_{\beta\alpha}(\gamma_\mu)_{\alpha\beta}$$

Therefore

$$\text{Trace } (\gamma_\mu \gamma_\nu) = \tfrac{1}{2} \text{ Trace}\{\gamma_\mu, \gamma_\nu\}.$$

Now note that the anti-commutator

$$\{\gamma_\mu, \gamma_\nu\}_{\alpha\beta} = 2g_{\mu\nu}(I_4)_{\alpha\beta}$$

where I_4 is the 4×4 unit matrix.

$\therefore \quad \text{Trace}\ \{\gamma_\mu, \gamma_\nu\} = 8g_{\mu\nu}.$

$\therefore \quad \text{Trace}\ (\gamma_\mu\gamma_\nu) = 4g_{\mu\nu}.$

A4.3 At 2 GeV CM energy $\sigma_{\mu\mu} = 4\pi\alpha^2/3s = 5{\cdot}58 \times 10^{-5}$. In order to convert to SI units we multiply by 0·389 mb; giving 21·7 nb. At 40 GeV CM energy this falls a factor 400 to 54·2 pb. 1 picobarn is 10^{-12} barns or 10^{-36} cm^2.

2 GeV is below the threshold for producing $\tau^+\tau^-$, so the cross-section is zero. The masses m_τ and m_μ are both small compared to 40 GeV, so $\sigma_{\tau\tau}$ is also 54·2 pb at 40 GeV.

A5.1 The effect of an infinitesimal transformation is

$$T(\delta a) = 1 + i\delta a p_x.$$

Then the operation produced on the x-coordinate is

$$\begin{aligned} x \rightarrow T^{-1}xT &= (1 - i\delta a p_x)x(1 + i\delta a p_x) \\ &= x - i\delta a[p_x, x]. \end{aligned}$$

Now, introducing the operator form of p_x, namely $-i\partial/\partial x$, we get

$$x \rightarrow x - \delta a,$$

which is a translation along the x-axis.

If all physical processes are invariant under translations $T(\delta a)$ commutes with every operator. Then p_x also commutes with every operator and so it is a conserved quantity.

A5.2 If the photon spins are aligned parallel (LR' or $L'R$) the spin component along $0z$ namely (S_z) is ± 2. On the other hand, LL' or RR' have S_z equal to zero. The total wavefunction of a photon travelling along the z-axis is $\psi = P\ \text{exp}i(\omega t - kz)$, where P describes the polarization state. We obtain the orbital angular momentum component along 0_z using

$$L_z\psi = i\partial\psi/\partial\phi,$$

where ϕ is the rotation angle around $0z$. Thus L_z is identically zero for a plane wave along its direction of motion. Now the total angular momentum component $J_z = L_z + S_z$, so $J_z = S_z$.

Evidently LR' or $L'R$ can be ruled out. Under the parity transformation

$$P(RR')=LL' \quad \text{and} \quad P(LL')=RR'$$

including a factor $+1$ from the intrinsic parity of a boson-antiboson pair. Thus $LL'\pm RR'$ has even (odd) parity. These combinations have the correct polarization states for the decay products from a scalar (pseudoscalar) parent.

A6.1

$$F_1+iF_2(=I^c_+)=\begin{bmatrix} 0 & 1 & 0 \\ 0 & 0 & 0 \\ 0 & 0 & 0 \end{bmatrix}$$

and

$$F_1-iF_2(=I^c_-)=\begin{bmatrix} 0 & 0 & 0 \\ 1 & 0 & 0 \\ 0 & 0 & 0 \end{bmatrix}$$

Thus $I^c_+ y=r$, $I^c_- r=y$ and all other operations give zeros. We also use the notation $V^c_\pm=(F_4\pm iF_5)$ and $U^c_\pm=(F_6\pm F_7)$. Then $V^c_+ b=r$, $V^c_- r=b$, $U^c_+ b=y$ and $U^c_- y=b$ with other operations annihilating the states (i.e. giving zero). I^c_+, etc., are the raising and lowering operators of $SU(3)_c$. In the $\bar{3}$ representation they should produce identical changes in I^c_3, etc. Now

$$\bar{r}=\begin{bmatrix} 1 \\ 0 \\ 0 \end{bmatrix}, \quad \bar{y}=\begin{bmatrix} 0 \\ 1 \\ 0 \end{bmatrix} \quad \text{and } \bar{b}=\begin{bmatrix} 0 \\ 0 \\ 1 \end{bmatrix}$$

while

$$\lambda_1=\begin{bmatrix} 0 & -1 & 0 \\ -1 & 0 & 0 \\ 0 & 0 & 0 \end{bmatrix}, \quad \lambda_2=\begin{bmatrix} 0 & -i & 0 \\ i & 0 & 0 \\ 0 & 0 & 0 \end{bmatrix},$$

$$\lambda_3 = \begin{bmatrix} -1 & 0 & 0 \\ 0 & 1 & 0 \\ 0 & 0 & 0 \end{bmatrix} \text{ etc. Then } I_+^c r = -\bar{y},$$

$$I_-^c \bar{y} = -\bar{r},\ I_3^c \bar{r} = -\tfrac{1}{2}\bar{r},\ I_3^c \bar{y} = \tfrac{1}{2}\bar{y}$$

etc. These results are consistent because (a) I_+^c raises I_3^c by one unit, (b) I_-^c lowers I_3^c by one unit and (c) $I_3^c(\bar{y}) = -I_3^c(y)$, etc.

Next we tackle the quark-antiquark states: $I_+^c(r\bar{r}) = -r\bar{y}$, $I_+^c(y\bar{y}) = r\bar{y}$ and $I_+^c(b\bar{b}) = 0$. Hence I_+^c annihilates $(r\bar{r} + y\bar{y} + b\bar{b})/\sqrt{2}$. Similarly all the other operators annihilate this state and hence it must be an SU(3) singlet with no partners.

A6.2 $$\alpha_s(q^2) = [B \ln(|q^2|/\Lambda^2)]^{-1}$$

where

$$B = (11 - 2N_F/3)/4\pi = 0{\cdot}557.$$

Therefore:

$$\alpha_s(10^6) = 1{\cdot}8/\ln(10^6/0{\cdot}04) = 0{\cdot}106$$

and

$$\alpha_s(10^{30}) = 1{\cdot}8/\ln(10^{30}/0{\cdot}04) = 0{\cdot}025.$$

At the higher energy the strong coupling constant has fallen to a value comparable to that of the electromagnetic fine-structure constant.

A7.1 (a) The quark content of K^0 is $d\bar{s}$ and of $\bar{K}^0$ is $\bar{d}s$. Then in the weak decay

$$(\bar{d})s \rightarrow (\bar{d})u + e^- + \bar{\nu}_e$$

or

$$\bar{K}^0 \rightarrow \pi^+ + e^- + \bar{\nu}_e.$$

Similarly in the weak decay

$$(d)\bar{s} \rightarrow (d)\bar{u} + e^+ + \nu_e$$

or

$$K^0 \rightarrow \pi^- + e^+ + \nu_e.$$

(b) For the decays of D^+ at quark level

$$c \rightarrow s' + W^+$$
$$\hookrightarrow u + \bar{d}'.$$

When the Cabibbo projection of s′ and d′ onto s and d is made the decay amplitude for

$$c \rightarrow s + u + \bar{d}$$

has a Cabibbo factor $\cos^2 \theta_C$ and hence an intensity proportional to $\cos^4 \theta_C$. On the other hand the decay amplitude for

$$c \rightarrow d + u + \bar{s}$$

has a Cabibbo factor $\sin^2 \theta_C$ and hence an intensity proportional to $\sin^4 \theta_C$. This indicates that $D^+ \rightarrow K^+ + \pi^+ + \pi^-$ is suppressed by a factor $\tan^4 \theta_C$ or about 2×10^{-3} with respect to $K^- + \pi^+ + \pi^+$.

A7.2 (a) calculated Fermi model cross-section is $G^2 s/\pi$. Then the unitary limit is reached when

$$G^2 s/\pi > 2\pi/s$$

i.e. when

$$s^2 > 2\pi^2/G^2$$

i.e. when

$$s > \sqrt{2}\pi/G \simeq 5 \times 10^5 \text{ GeV}$$

i.e. when

$$\sqrt{s} > 700 \text{ GeV}.$$

This gives a good idea of the energy scale at which electroweak unification should occur, namely well below 700 GeV.

(b) The CM energy squared

$$s=(E_\nu+m_e)^2-p_\nu^2$$

$$\simeq 2E_\nu m_e.$$

Therefore the momenta and energies of all the particles in the CM are $k^*=\sqrt{s/4}=\sqrt{E_\nu m_e/2}$ in the limit that masses are negligible. The parameters of the Lorentz boost from laboratory to CM are

$$\gamma=E_\nu/\sqrt{s}=\sqrt{E_\nu/2m_e} \text{ and } \beta\simeq 1.$$

Then the muon longitudinal momentum in the laboratory is

$$p_L=k^*(\cos\theta^*+1)\gamma$$

where θ^* is the CM scattering angle.

The muon transverse momentum in either frame is

$$p_T=k^*\sin\theta^*.$$

Thus the laboratory scattering angle is given by

$$\tan\theta=p_T/p_L=\sin\theta^*/\gamma(\cos\theta^*+1)$$

which is bounded by

$$\tan\theta(\max)=1/\gamma=\sqrt{2m_e/E_\nu}.$$

This limit is small compared to the angles when scattering occurs off a nucleus. An angle cut is therefore useful in making a reliable selection of scatters from electrons.

A8.1 In $\nu_e e^-$ scattering charged W exchange converts ν_e to e^- and e^- to ν_e. In $\bar{\nu}_e e^-$ scattering the incident fermions annihilate to give W^- which subsequently converts to $\bar{\nu}_e e^-$. When the incoming neutrino is of the μ variety these processes cannot occur because μ^-- and e-lepton number would not be conserved at any of the vertices.

A8.2 (a) The couplings of the Z^0 to each type of lepton and quark are given in Table 8.1. For a right-handed e^- $\sin^2\theta_W$; and for a left-handed $e^-(\sin^2\theta_W-\frac{1}{2})$. Thus the rate for Z^0 decay to

e^+e^- is $\Gamma(e^+e^-) \propto [(\sin^2\theta_W)^2 + (\sin^2\theta_W - \frac{1}{2})^2] = 2\sin^4\theta_W - \sin^2\theta_W + \frac{1}{4}$. The other rates are obtained in the same manner, with an additional colour factor of 3 for rates of decay to quarks.

(b) If we ignore kinematic factors which reduce the decay rate for decays to $t\bar{t}$ and to a lesser extent to $b\bar{b}$ then the total rate $\Gamma(\text{all}) \propto (24 - 48\sin^2\theta_W + 64\sin^4\theta_W)$. With $\sin^2\theta_W$ equal to 0·215, $\Gamma(\text{all}) \propto 16{\cdot}64$ including all three generations of quarks and leptons. Then

$$\Gamma(\nu_e\bar{\nu}_e)/\Gamma(\text{all}) = 0{\cdot}06 \text{ and } \Gamma(e^+e^+)/\Gamma(\text{all}) = 0{\cdot}03.$$

(c)

$$\Gamma(\text{all}) = (GM_z^3/12\pi\sqrt{2})/0{\cdot}06$$
$$= 2{\cdot}794 \text{ GeV}.$$

A9.1 The energy transfer ν is 40 GeV. The four-momentum transfer squared $q^2 = -2EE'(1-\cos\theta) = -9{\cdot}75(\text{GeV}/c^2)^2$. Then $x = -q^2/2M\nu = 0{\cdot}13$. From the data given in Fig. 9.3 the probability that the active parton is a quark is $0{\cdot}96/(0{\cdot}96+0{\cdot}18) = 0{\cdot}84$.

A9.2 The Drell-Yan cross-section is

$$d^2\sigma/dx_F\,dM = (8\pi K\alpha^2/9M^3)(\tau/\sqrt{x_F^2+4\tau})$$
$$\Sigma Q_f^2[q_f(x_1)\bar{q}_f(x_2) + \bar{q}_f(x_1)q_f(x_2)].$$

In the case that $x_F = 0$, the quark x-values are equal, $\tau = M^2/s = x^2$ and $x = x_1 = x_2$. When the pairs are massive, x is large and only the valence quarks are important. For these we have

$$q_f(x_1) = \bar{q}_f(x_2) = v_f(\sqrt{\tau})$$

where v_f is the valence quark (antiquark) distribution in the proton (antiproton). $\bar{q}_f(x_1)$ and $q_f(x_2)$ are negligible. Therefore:

$$d^2\sigma|dx_F\,dM|(x_F=0) = (4\pi K\alpha^2/9W^3\tau)\Sigma Q_f^2 v_f^2(\sqrt{\tau}).$$

A10.1 The valence $\bar{c}c$ pair making up the ψ' continue through the reaction unchanged to emerge as the valence $\bar{c}c$ pair making up the ψ. Then the two π-mesons must materialize from gluon emission by either the c-quark or $\bar{c}$-antiquark. If a diagram is drawn it will be of the Zweig suppressed form: some final-state particles are connected by gluon lines only to the remainder of the diagram.

A10.2 (a) The q_1q_1 quarks are identical quarks in an S-state. Their colour state is antisymmetric so that their space-spin-flavour wavefunction must be symmetric. Consequently they have parallel spins, giving a total spin of 1. Then using the Clebsch-Gordan coefficients for coupling the spin 1 of q_1q_1 to the spin $\frac{1}{2}$ of q_2 to give a total spin $|\frac{1}{2}, \frac{1}{2}\rangle$ for the baryon

$$|\tfrac{1}{2}, \tfrac{1}{2}\rangle = [\sqrt{2/3}\,|1, 1\rangle_1|\tfrac{1}{2}, -\tfrac{1}{2}\rangle_2 - \sqrt{1/3}\,|1, 0\rangle_1|\tfrac{1}{2}. +\tfrac{1}{2}\rangle_2].$$

In the $|1, 1\rangle\,|\frac{1}{2}, -\frac{1}{2}\rangle$ state the magnetic moments of q_1q_1 are aligned up and opposed by that of q_2. In the $|1, 0\rangle\,|\frac{1}{2}, +\frac{1}{2}\rangle$ state only the q_2 magnetic moment contributes. Thus

$$\mu(B) = (2/3)\,(2\mu(1) - \mu(2)) + (1/3)\mu(2).$$

(b) For the quarks:

$$\mu(\mathrm{u}) = g(2e/3)/2m_\mathrm{u} = 0{\cdot}67\mu$$
$$\mu(\mathrm{d}) = g(-e/3)/2m_\mathrm{d} = -0{\cdot}33\mu$$
$$\mu(\mathrm{s}) = g(-e/3)/2m_\mathrm{s} = -0{\cdot}23\mu$$

where the constituent masses are used. For the proton:

$$\mu_\mathrm{p} = (2/3)\,(2\mu(\mathrm{u}) - \mu(\mathrm{d})) + (1/3)\mu(\mathrm{d})$$
$$= 4/3\;\mu(\mathrm{u}) - \mu(\mathrm{d})/3 = \mu$$

so that μ is 2·79 nuclear magnetons.

(c) Now $\mu(\mathrm{u}) = 1{\cdot}86$, $\mu(\mathrm{d}) = -0{\cdot}93$ and $\mu(\mathrm{s}) = -0{\cdot}65$ n m. For the other baryons in the $\frac{1}{2}^+$ octet:

n(udd)	$-1{\cdot}86$	$(-1{\cdot}91)$
Σ^+(uus)	$+2.70$	$(+2{\cdot}38 \pm 0{\cdot}020)$
Σ^-(dds)	$-1{\cdot}02$	$(-1{\cdot}14 \pm 0{\cdot}05)$
Ξ^0(ssu)	$-1{\cdot}49$	$(-1{\cdot}25 \pm 0{\cdot}014)$
Ξ^-(ssd)	$-0{\cdot}56$	$(-0{\cdot}69 \pm 0{\cdot}04)$

where the values in brackets are the measured magnetic moments in n m.

(d) The Λ-hyperon quark content is uds. Both the Λ-hyperon and s-quark have isospin zero so that the (ud) quark pair must also have isospin zero. Thus the isospin (flavour) wavefunction of the (ud) pair is antisymmetric. Then to make the (ud) pair spin-space-flavour wavefunction overall symmetric the spin part must also be antisymmetric, i.e.

$$[u(\tfrac{1}{2})d(-\tfrac{1}{2})-u(-\tfrac{1}{2})d(+\tfrac{1}{2})]/\sqrt{2}$$

where the spin components appear within the brackets. The ud pair does not therefore contribute to the magnetic moment and $\mu_\Lambda=\mu(\mathrm{s})=-0{\cdot}65\ (-0{\cdot}613\pm0{\cdot}004)$ n m.

A10.3 Taking the second reaction $\pi^- + \mathrm{p} \rightarrow \Delta^0 \rightarrow \pi^0 + \mathrm{n}$, we have for the production step an isospin Clebsch-Gordan coefficient

$$(3/2,\ -\tfrac{1}{2}|1,\ -1;\ \tfrac{1}{2},\ +\tfrac{1}{2})=\sqrt{1/3}$$

and for the decay step

$$(3/2,\ -\tfrac{1}{2}|1,\ 0;\ \tfrac{1}{2}\ -\tfrac{1}{2})=\sqrt{2/3}.$$

The overall amplitude contains an isospin factor $\sqrt{2}/3$ and the intensity (rate) a factor 2/9. For the third reaction the isospin Clebsch-Gordan factors are both $1/\sqrt{3}$ and the intensity carries an isospin factor 1/9. In the first and last reactions the initial and final states have isospin 3/2 so the amplitude and rate factors are all unity.

A11.1 The definition of y is $\frac{1}{2}\ln[(E+p_\mathrm{L})/(E-p_\mathrm{L})]$. Hence

$$\mathrm{d}y/\mathrm{d}p_\mathrm{L}=[E-p_\mathrm{L}(\mathrm{d}E/\mathrm{d}p_\mathrm{L})]/(E^2-p_\mathrm{L}^2).$$

Also

$$\mathrm{d}E/\mathrm{d}p_\mathrm{L}=p_\mathrm{L}/E.$$

Therefore

$$\mathrm{d}y/\mathrm{d}p_\mathrm{L}=1/E.$$

Hence

$$N(p_L)dp_L \alpha \; dp_L/E = dy.$$

A11.2 In the forward direction $t \simeq 0$. The CM energy squared is given by $2m_p p_\pi$, where m_p is the proton mass and p_π the π-meson laboratory momentum in the kinematic range $p_\pi \gg m_p$. Thus at 20 GeV/c, s is 37·6 GeV and at 200 GeV/c, s is 376 GeV. The ratio of the cross-section at these CM energies is:

$$(d\sigma/dt)_{200}/(d\sigma/dt)_{20} = (376/37{\cdot}6)^{2\alpha(0)-2}$$
$$= 10^{-0{\cdot}9} = 0{\cdot}13.$$

A11.3 The quark content of the mesons involved is: $K^- \equiv (\bar{u}s)$, $\rho^0 \equiv (u\bar{u} - d\bar{d})/\sqrt{2}$ and $\omega \equiv (u\bar{u} + d\bar{d})/\sqrt{2}$ for the case of ideal mixing. Then in either reaction the s-quark is replaced by a u-quark. Now

$$\bar{u}s \rightarrow u\bar{u} = (u\bar{u} + d\bar{d})/2 + (u\bar{u} - d\bar{d})/2.$$

Thus the amplitudes for producing ρ^0 and ω are identical, and the rates are identical. Note that kinematic differences between the reactions are small because the masses of the $\rho(770)$ and the $\omega(783)$ are quite similar.

Appendix G
Tables of particle properties

These tables list masses, lifetimes, common and interesting decay modes and their branching fractions for the leptons and for the lightest hadrons of each species. The entries are drawn from the biennial 'Review of Particle Properties' currently to be found in *Physics Letters* **170B**, 1–350, April 1986. There the reader can find detailed information on the excited hadron states, i.e. those in which radial or orbital modes of the quark motion are excited. Whereas there is the single entry appearing below for the $\Lambda(1116)$, in the review this is accompanied by thirteen entries on well-established Λ-states having the same quark content and strong isospin; ranging from $\Lambda(1405)$ with $J^P = \frac{1}{2}^-$ to $\Lambda(2350)$ with $J^P = 9/2^+$. In the review mentioned there is a description of a revised convention for naming hadrons. The rules of this convention are listed below, following the tables of particles properties. Finally, please note that for simplicity the antineutrinos and neutrinos appearing in hadron decay modes are all written as ν.

Table G.1 *The leptons* ($J=\frac{1}{2}$)

Name	*Mass MeV/c²*	*Mean lifetime (sec.)*	*Decay modes and branching fractions*
ν_e	$<1{\cdot}0\times10^{-5}$	—	—
e^-	$0{\cdot}5110034$ $\pm0{\cdot}0000014$	$>2{\cdot}10^{22}$ years	—
ν_μ	$<0{\cdot}25$	—	—
μ^-	$105{\cdot}65916$ $\pm0{\cdot}00030$	$(2{\cdot}19703\pm0{\cdot}00004)$ $\times10^{-6}$	$e^-\bar{\nu}_e\nu_\mu$ (100%)
ν_τ	<70	—	—
τ^-	$1784{\cdot}2\pm3{\cdot}2$	$(3{\cdot}3\pm0{\cdot}4)\times10^{-13}$	$\mu^-\bar{\nu}_\mu\nu_\tau$ $(17{\cdot}6\pm0{\cdot}6)$% $e^-\bar{\nu}_e\nu_\tau$ $(17{\cdot}4\pm0{\cdot}5)$% $\pi^-\nu_\tau$ $(10{\cdot}1\pm1{\cdot}1)$% $\rho^-\nu_\tau$ $(21{\cdot}8\pm2{\cdot}0)$%

Table G.2 *The* s-*wave mesons* ($J^{PC}=0^{-+}$)

I	*Particle*	*Mass MeV/c²*	*Mean life or width*	*Decay modes and branching fractions*
1	π^+	139·57	$2{\cdot}60\times10^{-8}$ s	$\mu^+\nu \sim 1{\cdot}0$; $e^+\nu$ $1{\cdot}23\times10^{-4}$
	π^0	134·96	$0{\cdot}87\times10^{-16}$ s	$\gamma\gamma$ 98·8%
	π^-	139·57	$2{\cdot}6\times10^{-8}$ s	$\mu^-\nu \sim 1{\cdot}0$; $e^-\nu$ $1{\cdot}23\times10^{-4}$
0	η^0	548·8	1·05 KeV	$\pi^+\pi^0\pi^-$ 23·7%; $3\pi^0$ 31·92; $\gamma\gamma$ 38·9%
½	K^+	493·67	$1{\cdot}237\times10^{-8}$ s	$\mu^+\nu$ 63·5%; $\pi^+\pi^0$ 21·2%; $\pi^+\pi^+\pi^-$ 5·6%; $\pi^0e^+\nu$ 4·8%
	K^0	497·72	K_S $0{\cdot}8923\times10^{-10}$ s	$\pi^+\pi^-$ 68·6%; $\pi^0\pi^0$ 31·4%
			K^0_L $5{\cdot}183\times10^{-8}$ s	$\pi^0\pi^0\pi^0$ 21·5%; $\pi^+\pi^0\pi^-$ 12·4%; $\pi^\pm\mu^\mp\nu$ 27·1%; $\pi^\pm e^\mp\nu$ 38·7%
0	$\eta'(958)$	957·57	0·24 MeV/c^2	$\eta\pi\pi$ 65·2%; $\rho^0\gamma$ 30·0%
0	$\eta_c(2980)$	2981	11 MeV/c^2	$\eta'\pi\pi$ 4·1%; $K\bar{K}\pi$ 5·4%

Table G.2 (*continued*)

$\frac{1}{2}$	D^+	1869·3	$9{\cdot}2\times10^{-13}$ s	e^+ any 18·2%; $K^-\pi^+\pi^+$ 11·4% $K^-\pi^+\pi^+\pi^0$ 6·4%; $\bar{K}^0\pi^+\pi^0$ 13·4% $\bar{K}^0\pi^+\pi^+\pi^-$ 15·2%
	D^0	1864·6	$4{\cdot}3\times10^{-13}$ s	e^+ any 7%; K^- any 44%; $K^-\pi^+\pi^0$ 17·3%; $K^-\pi^+\pi^+\pi^-$ 10·9% $\bar{K}^0\pi^+\pi^-$ 8·5%
0	D^+ s	1970·5	$2{\cdot}8\times10^{-13}$ s	$\phi\pi^+$; $\phi\pi^+\pi^+\pi^-$
$\frac{1}{2}$	B^+	5271·2	$1{\cdot}4\times10^{-12}$ s	$\bar{D}\pi^+$ 1·1%; $D^{*-}(2010)\pi^+\pi^+$ 2·7%; ψ any 1%
	B^0	5275·2		$\bar{D}^0\pi^+\pi^-$ 7%; $D^{*-}(2010)\pi^+$ 1·7%

Table G.3 *The s-wave mesons* ($J^{PC} = 1^{--}$)

I	*Particle*	*Mass MeV/c²*	*Width MeV/c²*	*Decay modes and branching fractions*
1	$\rho(770)$	770·0	153·0	$\pi\pi \sim 1{\cdot}0$; e^+e^- 0·0045%
0	$\omega(783)$	782·6	9·8	$\pi^+\pi^-\pi^-$ 89·6%; $\pi^0\gamma$ 8·7% e^+e^- 0·0067%
0	$\phi(1020)$	1019·5	4·22	K^+K^- 49·5%; $K_s^0K_L^0$ 34·3% $\pi^+\pi^-\pi^0$ 14·9%; e^+e^- 0·031%
½	$K^{*+}(892)$ $K^{*0}(892)$	892·1	51·1	$K\pi \sim 1{\cdot}0$
½	$D^{*+}(2010)$	2010·1	<2·0	$D^0\pi^+$ 49%; $D^+\pi^0$ 34%; $D^+\gamma$ 17%
	$D^{*0}(2010)$	2007·2	<5	$D^0\pi^0$ 51·5%; $D^0\gamma$ 48·5%
0	$\psi(3097)$	3096·9	0·063	e^+e^- 6·9%; hadrons + radiative 86·2%; $\mu^+\mu^-$ 6·9%;
0	$\Upsilon(9460)$	9460·0	0·043	e^+e^- 2·8%; $\tau^+\tau^-$ 3·2%; $\mu^+\mu^-$ 2·8%

Table G.4 *The* s-*wave baryons* ($J^P = \frac{1}{2}^+$)

I	*Particle*	*Mass MeV/*c^2	*Mean life sec*	*Decay modes and branching fractions*
$\frac{1}{2}$	p	938·28	—	—
	n	939·57	898 ± 16	$pe^-\nu$ 100%
0	Λ	1115·60	$2{\cdot}632 \times 10^{-10}$	$p\pi^-$ 64·2%; $n\pi^0$ 35·8%; $pe^-\nu$ $8{\cdot}3 \times 10^{-4}$
1	Σ^+	1189·37	$0{\cdot}800 \times 10^{-10}$	$p\pi^0$ 51·64%; $n\pi^+$ 48·36%; $\Lambda e^+\nu$ 2×10^{-5}
	Σ^0	1192·46	$5{\cdot}8 \times 10^{-20}$	$\Lambda\gamma \sim 100\%$
	Σ^-	1197·34	$1{\cdot}482 \times 10^{-10}$	$n\pi^- \sim 100\%$; $ne^-\nu$ $1{\cdot}02 \times 10^{-3}$; $\Lambda e^-\nu$ $5{\cdot}74 \times 10^{-5}$
$\frac{1}{2}$	Ξ^0	1314·9	$2{\cdot}90 \times 10^{-10}$	$\Lambda\pi^0 \sim 100\%$
	Ξ^-	1321·32	$1{\cdot}642 \times 10^{-10}$	$\Lambda\pi^- \sim 100\%$; $\Lambda e^-\nu$ $5{\cdot}5 \times 10^{-4}$
0	Λ_c^+	2281·2	$2{\cdot}3 \times 10^{-13}$	$pK^-\pi^+ 2{\cdot}2\%$; Λ any $\sim 33\%$

Table G.5 *The* s-*wave baryons* ($J^P = \frac{3}{2}^+$)

I	*Particle*	*Mass MeV*/c^2	*Width MeV*/c^2	*Decay modes and branching fractions*
$\frac{3}{2}$	$\Delta(1232)$	1232·0	115	$N\pi$ 99·4%; $N\gamma$ 0·6%
1	$\Sigma(1385)^+$	1382·8	36	$\Lambda\pi$ 88%; $\Sigma\pi$ 12%
	$\Sigma(1385)^0$	1383·7	36	
	$\Sigma(1385)^-$	1387·2	39	
$\frac{1}{2}$	$\Xi(1530)^0$	1531·8	9·1	$\Xi\pi$ 100%
	$\Xi(1530)^-$	1535·0	10·1	
0	Ω^+	1672·4	$0{\cdot}822 \times 10^{-10}$ s (lifetime)	ΛK^- 67·8%; $\Xi^0\pi^-$ 23·6% $\Xi^-\pi^0$ 8·6%; $\Xi^0 e^- \nu$ $5{\cdot}6 \times 10^{-3}$

Table G.6 *Conventions for naming hadrons*

1. *Mesons with net heavy flavour* (*c*, *b*, *t*) *equal to zero*

Quark content	J^{PC} *even*$^{+-}$	*odd*$^{-+}$	J^{--}	J^{++}
$u\bar{d}$, $(d\bar{d}-u\bar{u})$, $d\bar{u}$ $[I=1]$	π	b	ρ	a
$(u\bar{u}+d\bar{d})[I=0]$ } {	η	h	ω	f
$s\bar{s}$ } {	η'	h′	ϕ	f′
$c\bar{c}$	η_c	h_c	ψ	χ
$b\bar{b}$	η_b	h_b	Υ	χ_b

2. *Mesons with non-zero heavy flavour*

(a) If the heavy quark in the meson is s, c, b or t the symbol to be used is $\bar{K}$, D, $\bar{B}$ or T respectively. If it is the antiquark that is heavy the symbol chosen is K, $\bar{D}$, B or $\bar{T}$ respectively.
(b) If the lighter quark is not u or d, but 'x' then a subscript x is added.
(c) If the spin-parity lies in the series 0^+, 1^-, 2^+ ... a superscript * is added.
(d) The spin is added as a subscript unless the meson is a pseudoscalar or vector (0^-, 1^-).

3. *Baryons not carrying heavy flavour* (*c*, *b*, *t*)

Quark content	*Strong isospin*	*Symbol*
uud, udd	$\frac{1}{2}$	N
uuu, uud, udd, ddd	$\frac{3}{2}$	Δ
uds	0	Λ
uus, uds, dds	1	Σ
uss, dss	$\frac{1}{2}$	Ξ
sss	0	Ω

If the s-quark is replaced by a heavier quark this is indicated by a subscript. Thus Λ_c is an $I=0$ udc state while Ω_{ccc}^{++} is an $I=0$ ccc state.

Bibliography

Chapter 2

Anderson, C. D., *Phys. Rev.* **43**, 491 (1933).

Anderson, C. D. and Neddermayer, S., *Phys. Rev.* **51**, 884 (1937).

Barnes, V., *et al.*, *Phys. Rev. Lett.* **12**, 204 (1964).

Breidenbach, M., *et al.*, *Phys. Rev. Lett.* **23**, 935 (1969).

Brown, R., *et al.*, *Nature* **163**, 47 (1949).

Conversi, M., Pancini, E. and Piccioni, O., *Phys. Rev.* **71**, 209 (1947).

Danby, G., *et al.*, *Phys. Rev. Lett.* **9**, 36 (1962).

Farley, F. M. and Picasso, E., *Ann. Rev. Nucl. Sci.* **29**, 243 (1979).

Gell-Mann, M., *Phys. Lett.* **8**, 214 (1964).

Lattes, C. M. G., Muirhead, G. H., Powell, C. F. and Occhialini, G. P., *Nature* **159**, 694 (1947).

Pauli, W., *Handbuch der Physik* **24**, 1, 233 (1933).

Perl, M. L. *et al.*, *Phys. Rev. Lett.* **35**, 1489 (1975).

Reines, F. and Cowan, C., *Phys. Rev.* **113**, 273 (1959).

Van Dyck, R. S., Schwinberg, P. B. and Dehmelt, H. G., *Phys. Rev. Lett.* **38**, 310 (1977).

Zweig, G., CERN Report 8419/Th412 (1964).

General

Close, F., *An Introduction to Quarks and Partons*, Academic Press, London, 1979.

Perkins, D. H., *Introduction to High Energy Physics*, 2nd edn, Addison-Wesley, Reading, Mass., 1982.

Rossi, B., *High Energy Particles*, Prentice-Hall, New York, 1952.

Chapter 3

Arnison, G., *et al., Phys. Lett.* **122B**, 103 (1983).
Barnes, V., *et al., Phys. Rev. Lett.* **12**, 204 (1964).
Charpak, *et al., Nucl. Inst. Meth.* **62**, 262 (1968).
Gluckstern, R. L., *Nucl. Inst. Meth.* **24**, 381 (1963).
Kubota, Y. F., *et al., Nucl. Inst. Meth.* **217**, 249 (1983).
Pullia, A., *Riv. Nuovo Cim.* **7**, 2 (1984).

General

Kleinknecht, *Detectors for Particle Radiation*, CUP, Cambridge, 1986.
Particle Data Group, *Phys. Lett.* **170B**, 42–51 (1986).

Chapter 4

Borie, E. and Rinker, G. A., *Rev. Mod. Phys.* **54**, 67 (1982).
Feynman, R. P., *Rev. Mod. Phys.* **20**, 267 (1948).
Hanson, G., *et al., Phys. Rev. Lett.* **35**, 1609 (1975).
Lamb Jr, W. E. and Rutherford, R. C., *Phys. Rev.* **72**, 241 (1947).
Schwinger, J., *Phys. Rev.* **73**, 416 (1948).
Tomonaga, S., *Phys. Rev.* **74**, 224 (1948).

Chapter 5

Cartwright, J., *et al., Phys. Rev.* **91**, 667 (1953).
Emmerson, J. M., *Symmetry Principles in Particle Physics*, Clarendon, Oxford, 1972.
Lee, T. D. and Yang, C. N., *Phys. Rev.* **104**, 254 (1956).
Werner, S. A., Colella, R., Overhauser, A. W. and Eagan, C. F., *Phys. Rev.* **35**, 1053 (1975).
Wu, C. S., Ambler, E., Hayward, R., Hoppes, D. and Hudson, R., *Phys. Rev.* **105**, 1413 (1957).

Chapter 6

Arnison, G. *et al., Phys. Lett.* **123B**, 115 (1983).
Arnison, G. *et al., Phys. Lett.* **136B**, 294 (1984).
Arnison, G. *et al., Phys. Lett.* **158B**, 494 (1985).
Atherton, H. W. *et al., Phys. Lett.* **158B**, 81 (1985).
Banner, X. *et al., Phys. Lett.* **118B**, 203 (1982).

Barber, D. P. *et al.*, *Phys. Rev. Lett.* **43**, 830 (1979).
Bartel, W. *et al.*, *Phys. Lett.* **119B**, 239 (1982).
Behrend, H. J. *et al.*, *Phys. Lett.* **144B**, 297 (1984).
Berger, C. *et al.*, *Phys. Lett.* **86B**, 418 (1979).
Brandelik, R. *et al.*, *Phys. Lett.* **83B**, 261 (1979).
Gell-Mann, M., *Phys. Lett.* **8**, 214 (1964).
Greenberg, O. W., *Phys. Rev. Lett.* **13**, 598 (1964).
Gross, D. J. and Wilczek, F., *Phys. Rev.* **D8**, 3633 (1973).
Politzer, H. D., *Phys. Rev. Lett.* **30**, 1346 (1973).

Chapter 7

Cabibbo, N., *Phys. Rev. Lett.* **10**, 531 (1963).
Chodos, A. *et al.*, *Phys. Rev.* **D9**, 3471, **D10**, 2599 (1974).
Christiansen, J. H., Cronin, J. W., Fitch, V. I. and Turlay, R., *Phys. Rev. Lett.* **13**, 138 (1964).
Fermi, E., *Z. Phys.* **88**, 161 (1934).
Gjesdal, S. *et al.*, *Phys. Lett.* **52B**, 113 (1974).
Kobayashi, M. and Maskawa, T., *Prog. Theor. Phys.* **49**, 652 (1973).

Chapter 8

Arnison, G. *et al.*, *Phys. Lett.* **122B**, 103, **129B**, 273, **126B**, 398 (1983).
Bergsma, F. *et al.*, *Phys. Lett.* **117B**, 272 (1983).
Glashow, S., *Nucl. Phys.* **22**, 579 (1961).
Glashow, S., Iliopoulos, J. and Maiani, L., *Phys. Rev.* **D12**, 1285 (1970).
Goldstone, J., Salam, A. and Weinberg, S., *Phys. Rev.* **127**, 965 (1962).
Hasert, F. J. *et al.*, *Phys. Lett.* **46B**, 121 (1973a).
Hasert, F. J. *et al.*, *Phys. Lett.* **46B**, 138 (1973b).
Higgs, P. W., *Phys. Rev. Lett.* **12**, 132, **13**, 509 (1964).
Kim, J. E. *et al.*, *Rev. Mod. Phys.* **53**, 211 (1981).
Rubbia, C., McIntyre, P. and Cline, D., *Proc. 1976 Int. Neutrino Conf.*, Aachen, Braunschweig, 1977, p. 683.
Salam, A., *Elementary Particle Theory: Relativistic Groups and Analyticity*, Nobel Symposium No. 8 (edited by N. Svartholm), Almquist & Wiksell, Stockholm, 1968.

Van der Meer, S., *Stochastic Damping of Betatron Oscillations in the ISR*, CERN/ISR-PO/172-13 (1972).
Weinberg, S., *Phys. Rev. Lett.* **19**, 1264 (1967).

Chapter 9

Altarelli, R. and Parisi, G., *Nucl. Phys.* **B126**, 298 (1977).
Bergsma, F. *et al.*, *Phys. Lett.* **123B**, 269 (1983).
Bodek, A. *et al.*, *Phys. Rev.* **D20**, 1471 (1980).
Demioz, M., Ferroni, F. and Longo, E., *Phys. Rep.* **130**, 293 (1986).
Drell, S. D. and Yan, T-M., *Phys. Rev. Lett.* **24**, 181 (1970).
de Groot, J. G. H. *et al.*, *Z. Phys.* **C1**, 143 (1979).
Kenyon, I. R., *Rep. Prog. Phys.* **45**, 1213 (1982).

Chapter 10

Aguilar-Benitez, M. *et al.*, *Rev. Mod. Phys.* **56**, S1 (1984).
Buchmuller, W. and Tye, S. H., *Phys. Rev.* **D24**, 132 (1981).
Gasser, J. and Leutwyler, H., *Phys. Rep.* **87**, 78 (1982).
Gell-Mann, M., *Phys. Lett.* **8**, 214 (1964).
de Rujula, A., Giorgi, H. and Glashow, S. L., *Phys. Rev.* **D12**, 147 (1975).
Williams, W. S. C., *An Introduction to Elementary Particles*, 2nd edn, Academic Press, London, 1971.
Zweig, G., CERN Report 8419/Th 412 (1964).

Chapter 11

Aguilar-Benitez, *et al.*, *Rev. Mod. Phys.* **56** (1984).
Arnison, G. *et al.*, *Phys. Lett.* **136B**, 294 (1984).
Arnison, G. *et al.*, CERN-EP/86-55 (May 1986).
Barnes, A. *et al.*, *Phys. Rev. Lett.* **37**, 76 (1976).
Ciapetti, *et al.*, *Nucl. Phys.* **B64**, 58 (1973).
Cocconi, G., *Phys. Lett.* **49B**, 459 (1974).
Combridge, B. L., Kripfganz, J. and Ranft, J., *Phys. Lett.* **70B**, 234 (1977).
Drijard, D. *et al.*, *Nucl. Phys.* **B155**, 269 (1979).
Feynman, R. P., *Phys. Rev. Lett.* **23**, 1415 (1969).
Hanbury-Brown, R. and Twiss, R. Q., *Phil. Mag.* **45**, 663 (1954), *Nature* **178**, 1046 (1956).

Irving, A. C. and Worden, R. P., *Phys. Rep.* **C34**, 118 (1977).
Jacob, M., 'Collider Physics—Present and Prospects', SLAC Summer Institute (1983).
Regge, T., *Nuovo Cim.* **14**, 951 (1959).

Chapter 12

Callaway, D. J. E., *Cont. Phys.* **26**, 23 (1985).
Frampton, P. H. and Vogel, P., *Phys. Rep.* **82**, 339 (1982).
Misner, C. W., Thorne, K. S. and Wheeler, J. A., *Gravitation*, Freeman, San Francisco, 1970.
Perkins, D. H., *Ann. Rev. Nucl. Part. Sci.* **34**, 1 (1984).
Ramond, P., *Ann. Rev. Nucl. Part. Sci.* **33**, 31 (1983).
Sakarov, A. D., *JEPT Lett.* **5**, 24 (1967).
Schwartz, J. H., *Comm. Nucl. Part. Phys.* **13**, 103 (1984), **15**, 9 (1985).
Vergados, J. D., *Phys. Rep.* **133**, 1 (1986).
Wess, J. and Zumino, B., *Phys. Lett.* **49B**, 52 (1974).

Index

Index